Mathographics

by Robert Dixon

DOVER PUBLICATIONS, INC., New York

This Dover edition, first published in 1991, is an unabridged, unaltered republication of the work first published by Basil Blackwell, Oxford, England, 1987.
This edition is published by special arrangement with Basil Blackwell Limited, 108 Cowley Road, Oxford OX4 1JF, England.

Manufactured in the United States of America
Dover Publications, Inc., 31 East 2nd Street, Mineola, N.Y. 11501

Library of Congress Cataloging-in-Publication Data

Dixon, Robert (Robert A.)
Mathographics / by Robert Dixon.
p. cm.
"First published in Great Britain in 1987 by B. Blackwell"—T.p. verso.
Includes index.
ISBN 0-486-26639-7
1. Mathematics—Data processing. 2. Computer graphics. I. Title.
QA76.95.D58 1991
510′.285′66—dc20 90-48830
CIP

Contents

List of drawings iv
Preface vi
How to Use this Book viii

1 Compass Drawings 1
- 1.1 Eggs for beginners 3
- 1.2 Euclidean constructions 12
- 1.3 Euclidean approximations 34
- 1.4 Gauss extends Euclid 52
- 1.5 The eight centres of a triangle 55
- 1.6 Inverse points and mid-circles 62

2 String Drawings 75

3 Perspective Drawings 79

4 The Story of Trigonometry 89
- 4.1 The theorem of Pythagoras 92
- 4.2 Square roots 96
- 4.3 The story of pi (π) 98
- 4.4 The story of sine and cosine 102

5 Computer Drawings 107
- 5.1 On drawing a daisy 122
- 5.2 Drawing in perspective 144
- 5.3 Ten elementary transformations 147
- 5.4 The fractal universe 169

Bibliography 207

Answers 208

Index 211

List of drawings

1 Moss' egg
2 Thom's ovals or eggs
3 $\sqrt{3}:2:\sqrt{7}$
4 Golden egg
5 Four-point egg
6 Five-point egg
7 Hexagon
8 Square
9 Regular pentagon
10 Regular decagon
11 Angle bisection
12 Right angles
13 Tangents to a circle
14 Line bisection
15 Line division by a given ratio
16 The square root of a number (5)
17 Golden ratio
18 Equilateral triangle
19 Heptagon: method I
20 Heptagon: method II
21 Heptagon: method III
22 Enneagon: method I
23 Enneagon: method II, version 1
24 Enneagon: method II, version 2
25 Rectifying the circle: method I
26 Rectifying the circle: method II
27 Squaring the circle: extension of method I
28 17-gon
29 Incircle and excircles of a triangle
30 Intersecting set of coaxal circles
31 Non-intersecting set of coaxal circles
32 Orthogonal sets of coaxal circles
33 String-drawn egg
34 Hexagon, 12-gon, 24-gon, 48-gon, 96-gon
35 Sine and cosine
36 π
37 Touching stars in concentric rings
38 Four views of an egg
39 Venus
40 Plughole vortex
41 Bubbles
42 Curves of collineation (after Lawrence Edwards)
43 Sphere
44 Crinkles
45 Waves
46 Ocean view
47 The complete Fermat spiral: $r = \pm\sqrt{a}$
48 Powers of x
49 Polygons, a circle and a sine wave

50 Archimedes' spiral: $r = ka$
51 Equiangular spiral: $r = k^a$
52 Fermat's spiral: $r = \sqrt{(ka)}$
53 Cylindrical phyllotaxis
54 Cylindrical phyllotaxis, with vertical compression
55 Spherical phyllotaxis
56 Exponential growth
57 Logistic growth
58 Sunflower
59 False daisies: divergence of $360°/C$, with $C = \pi$ and e
60 False daisies: $C = \sqrt{5}$ and 1.309
61 True daisy: $C = \tau$
62 Rose curves with a twist
63 Plane in perspective
64 Sphere in perspective
65 Translation
66 Rotation and translation
67 Chessboard reflected in a sphere
68 Inversion
69 Men on a plane
70 Men on a sphere
71 Spherical symmetry
72 Inversion of circles between parallel lines
73 Inversion of a chessboard about its centre
74 Ellipse inverted about external point on its major axis
75 Row of ellipses after an antiMercator
76 Rows of identical stars after an antiMercator
77 Inversion of Drawing **76**
78 Rotated view of Drawing **77**
79 Sphere
80 Drawing **79** inverted into a plane
81 Fractal tetrahedron (becoming a dust)
82 Rough surface
83 Nested circles
84 Fractal phyllotaxis: cauliflower
85 Blancmange curve
86 Koch curve: displacements alternating each generation
87 Dragon curve: displacements alternating each segment
88 Dragon curve
89 C-shaped curve: displacements all in the same direction of $\frac{1}{2}\sqrt{2}$
90 C-shaped curve
91 Dirichlet's function
92 Dürer's pentagons
93 Pentasnow
94 Non-overlapping circles randomly distributed
95 Condensation
96 Condensation
97 Craters in the plane or spheres in space
98 Moon
99 Moon view
100 Ripple
101 Ripple reflected
102 Ripples
103 Ripples
104 Ripples
105 Waves
106 Turtle geometry

Preface

In addition to the more usual activities of doing sums, solving equations and proving theorems, mathematics can also be about doing drawings. In this book a collection of drawing exercises is presented. The exercises come from two main areas of mathematical drawing — **compass constructions** and **computer graphics** — which complement and contrast with each other nicely. The first is an extremely ancient art, requiring no written skills at all but only the care that it takes to place a compass point accurately. The second could hardly be more modern and requires no manual skills. But you do need to have learnt to write a few simple programs and to know some basic school mathematics — mainly algebra and trigonometry.

In order to give a complete picture of the development of geometry, I have included a brief mention of **string drawings** and **perspective drawings**. However, other areas of mathematical drawing such as graphs and charts, linkages, bipolar curves, envelope constructions and so on, are not included. Nor do I deal with the many interesting topics in computer graphics beyond simple line drawing (vector graphics) such as colouring, shading, reflections, hidden line, animation and so on.

The book is aimed at a wide range of ages and abilities. For those who may have forgotten, Chapter 4 provides a summary of Pythagoras' theorem, square roots, π and sines and cosines to be used in Chapter 5. I have also supplied additional exercises (with answers) at the ends of the chapters, together with recommended further reading, detailed in the bibliography.

Although the book progresses from ancient to modern in a gradual manner, and with a steady accumulation of knowledge, it can also be read out of sequence as a collection of more or less individual topics. The aim has been to cover sufficient basic principles (of number, length and angle) to enable readers to carry out their own creative investigations. The artwork ranges from the totally abstract to the somewhat pictorial. Nature, as well as mathematics, has been my inspiration.

According to legend, it was Pythagoras who discovered that musical harmony answers to simple rules of whole numbers and who jumped to the conclusion that the whole of nature might likewise be explained. The origins of modern science and mathematics go back to this realisation. Leonardo da Vinci pondered many of nature's patterns in time and space, including the branching of trees and the movements of water waves. The idea that the look of the world, and also its beauty, might be studied mathematically has become established in this century under the headings of **morphology** (the shape of objects) and **morphogenesis** (the origin and growth of forms). Many of my topics are studied from this viewpoint.

Mathematical drawing must be as much concerned with craft (control, care and precision) as with mathematics. This is largely determined by the instruments used — it is important to use good instruments. For advice on compasses, turn to p. 1. For information on computing hardware and program languages used, see p. 107.

Finally, I should like to take this opportunity to thank all those who have helped and inspired this project, including the staff of *New Scientist*, staff and students at the Royal College of Art, and staff and students at the

City and East London College. I should like to mention in particular Professor H.S.M. Coxeter, the late Frank Malina, Dr Ensor Holiday, Christine Sutton, John Vince, Brian Reffin-Smith, Roger Malina, John Fife, Alan Senior, Don Manley, Ray Hemmings, Nick Mellersh, Debora Coombs, Stephanie Moss and Robert Oldershaw.

Robert Dixon
1987

ACKNOWLEDGEMENTS

British Museum (Natural History), London p. 130, Fig. 5.1C; Department of Archaeology, Royal Museum of Scotland p. 130, Fig. 5.1B.

How to Use this Book

Look at the drawings in this book before reading the text. They are the main contents. Try to work out the rules for constructing them. Draw them for yourself. Try to imagine ways of using these same rules to make other and perhaps totally different drawings of your own.

The text is designed to explain the theory behind the drawings but not to remove all opportunities for the reader to puzzle out the finer details and working practicalities. Puzzlement is generally acknowledged to be a good way to become involved with mathematics, whatever the level. The topics covered range from simple ideas to quite sophisticated notions which may take a little time to fathom out. Readers who feel 'mathophobic' can omit the algebra and trigonometry and yet still follow most of this book, which is about lines and circles and how to put them together to make various patterns, both abstract and natural.

The book begins with the classical use of compass and rule and proceeds via a lengthy interlude to the present use of the computer. These two contrasting mathematical drawing instruments are given roughly equal emphasis. You may be particularly interested in one or other of them, but I hope that you will consider the ideas which run from one to the other. In the interlude, string drawing and perspective drawing are briefly considered; this is followed by a set of wholly theoretical sections in which trigonometry is dealt with for the benefit of readers not acquainted with it.

If you read this book from cover to cover, you will find that the numerical sequence of chapters involves several deviations. Since it is not essential to pursue all of these, you are advised to find your way around this book by making use of the diagram (opposite) showing the structure of the main ideas. From this, as you may note, there are five main places in the book where you might profitably begin. The arrows indicate various sequences for reading the chapters. Readers wishing to follow the main theme of ideas in this book should work their way through the vertical sequence of eleven sections.

This book can be approached in a number of ways. The diagram suggests various sequences in which to approach the chapters.

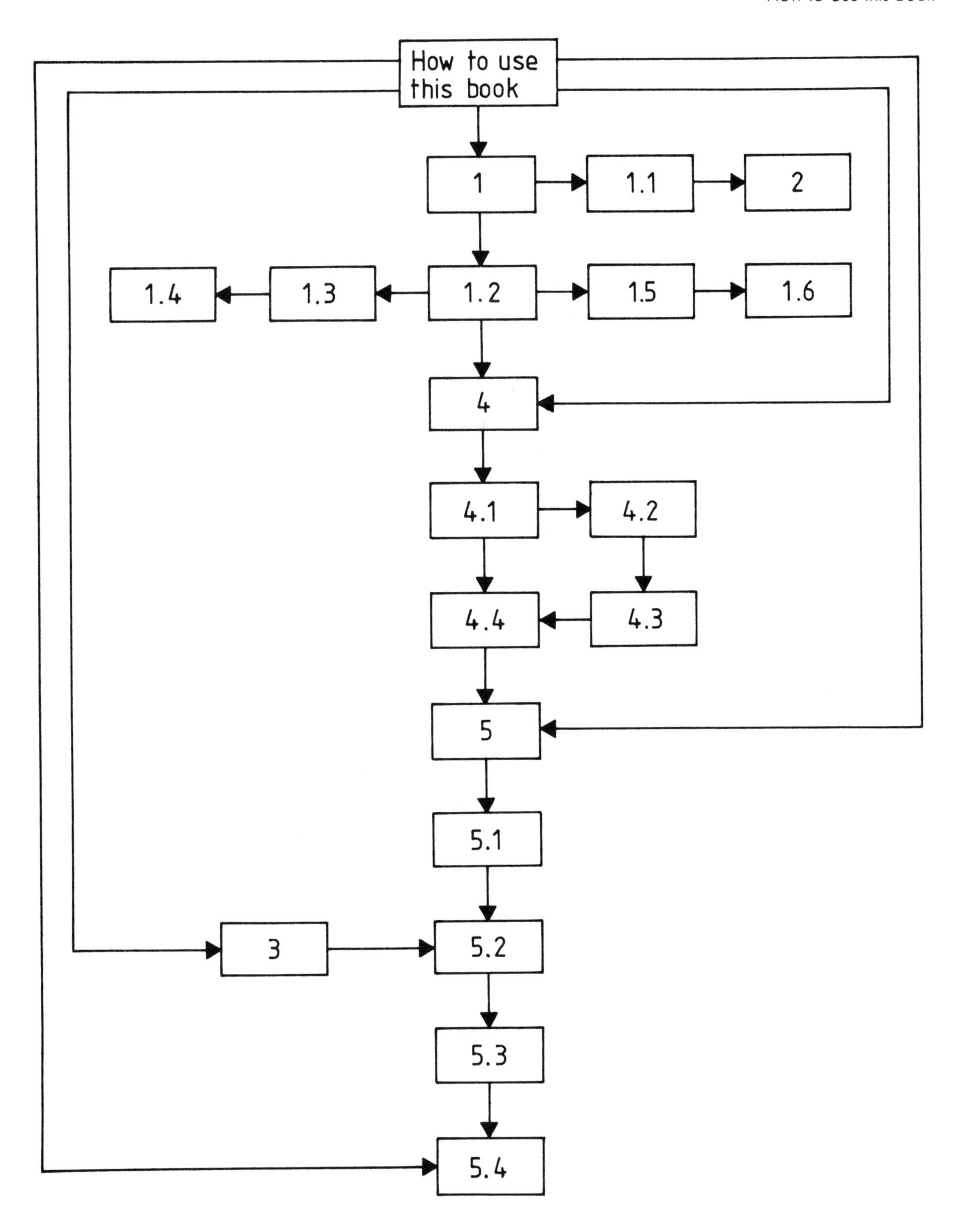
How to use this book
1
1.1
2
1.4
1.3
1.2
1.5
1.6
4
4.1
4.2
4.4
4.3
5
5.1
3
5.2
5.3
5.4

1 Compass Drawings

RULE AND COMPASS

The shortest distance between two points is a straight line — the line of sight. The ancient Egyptians used a stretched cord to form a line and also to define a length. The free end of a stretched cord whose other end is fixed can be used to trace a circle. We shall use a rule for drawing lines and a good modern compass for drawing circles.

Almost any wooden, plastic or metal rule will do, but check its straightness from time to time for warp or wear. If you intend to use a pen, you should avert the blotching effects of capillary action by chamfering a fraction of a millimetre off the drawing edge of the rule which makes contact with the paper. Hold a piece of sandpaper against this edge at a 45° angle — a few even sanding strokes will suffice. Alternatively, attach a single length of masking tape to the underside of the rule to create the required clearance. When drawing a line, hold the pen vertical throughout its path, and hold the rule firmly by forming a wide spread of fingers and thumb.

CHOOSING A COMPASS

Be warned; many compasses on sale are worse than useless, because they fail to maintain a constant distance between point and pen. The chief offenders are, of course, the cheaper instruments. If your first experiences of this art are accompanied by failure due to this fault, you will be lucky not to be put off further attempts for life. The remedy is simple; buy a good compass (Figure 1.0A). Even the most expensive compass costs less than an average pocket calculator. The recommended type should have the following features:

- a screw-thread radius adjustment
- hinged feet for obtaining vertical pen and point
- a rigid pen-holding attachment.

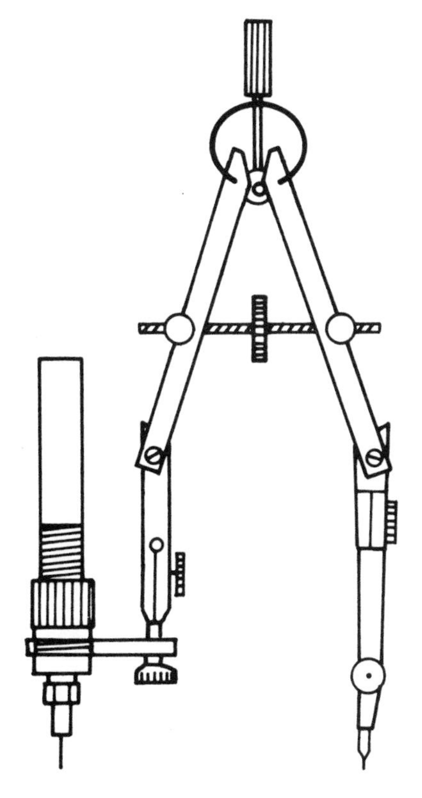

Figure 1.0A

THE RULES

In a compass drawing, all measurements are taken with the compass — the distance between point and pen. The graduations on your rule play no part in it. It is as if the inch, the centimetre and all other standard units of length did not exist.

A point is drawn by the intersection of two lines, two arcs, or a line and an arc. We shall also allow a well-made compass point hole.

Through any given point, it is possible to draw an infinite number of lines each with a different direction; infinitely many circles could be centred there, each with a different radius. Through any two points there is only one line. All lines can be extended indefinitely. A length is the distance between two points.

THE ERRORS

A typical compass drawing starts with an arbitrary line against which all subsequent directions will be measured, and an arbitrary length (radius) against which all subsequent lengths are measured. The step-by-step sequence of taking lengths and drawing arcs to find further points leads to an accumulation of errors. For this reason, there are practical limits to the number of steps which may be attempted. It also means that every effort should be made to minimise the error at each step. The most common source of error arises in positioning the compass point; the difference between the edge and the middle of even a fine pen line, for example, is a significant error. Intersections between arcs and/or lines intended to define points become unsatisfactory for small angles of intersection — less than 10°, say.

1.1 Eggs for beginners

CONSTRUCTING AN OVAL

A new exercise with compass and rule has appeared recently: constructing ovals. The basic idea is to form a smooth curve by joining several circular arcs of different radii. There must be no bend in the curve at the point where two arcs meet. How is it done?

The answer is that a straight line drawn through the centres of the two arcs must pass through the point where they join. Apart from that, the radii can be any pair of sizes and can even point in opposite directions. Figure 1.1A illustrates this principle and some simple uses of it.

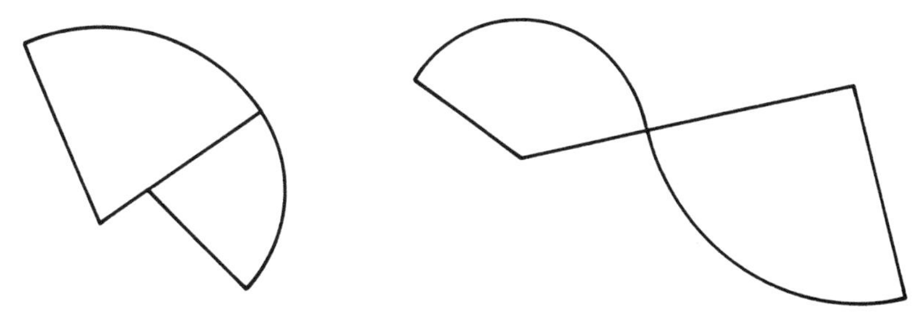

Figure 1.1A

Architects and engineers have long used this principle for constructing curves such as an 'elliptical' arch or an oval cam. Albrecht Dürer (1471–1528) used the principle to design a Roman alphabet, suitable for use by the mason or in the newly invented printing press. The geometric art of constructing ovals by this method is thought by some people to be twice as ancient as Euclid, dating from prehistoric times when Easter honoured Eostre, the goddess of springtime renewal.

For simplicity, this section is devoted to the single puzzle of constructing an oval which is approximately the shape of the outline of a bird's egg. There is no right solution, so far as we know, only a range and variety of more or less successful-looking curves. The lessons learnt about drawing egg shapes help to teach us about the ideas of Euclid. But the main lesson of this chapter is that *any* curve (and not just egg shapes) may be made up from a number of parts, each of which is part of a true circle.

THE SHAPE OF A BIRD'S EGG

Snails, insects, fish, reptiles and birds all lay eggs, but the shape of a bird's egg is particularly striking because of its geometric regularity. It is an example of what mathematicians call a **surface of revolution**. To see what this means, draw the outline of a bird's egg on a piece of paper. Draw another line through the centre of it. Then turn the piece of paper about the centre-line. As you turn the paper (as the paper *revolves*), the outline that you have drawn will trace out the complete surface of a bird's egg in space. Another property of a bird's egg is that the curvature is very smooth; it is almost as simple as a perfect sphere, but not quite. Apart from the elongation of its shape (which it shares with ellipses and ellipsoids), there is the marked asymmetry which we specifically call 'egg shaped', i.e. one end is blunter than the other.

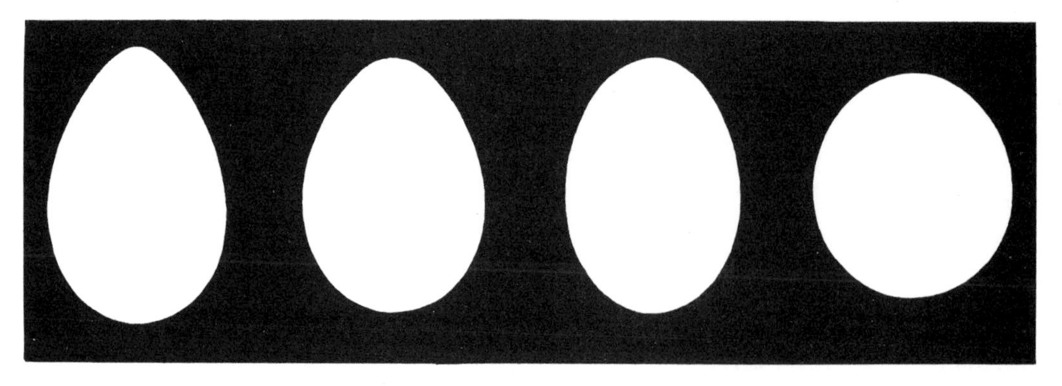

Figure 1.1B

Figure 1.1B shows a variety of birds' eggs from different species. As you can see, the shapes vary a great deal, but the general properties described above apply to all of them. So we need not worry if our attempts to construct an egg shape by compass and rule result in a similar variety of solutions. We shall obtain the basic egg-shaped effect by ensuring that the two ends have different radii of curvature.

WHY IS A BIRD'S EGG THE SHAPE THAT IT IS?

To this day, we do not know why birds' eggs are more pointed at one end than at the other. The suggestions made (that it prevents them from rolling off cliffs or that it provides an ideal packing solution for the clutch of a brooding hen) have never been proved. However, let us briefly look at the science behind naturally occurring curves such as this.

First of all, the shell of a bird's egg is highly brittle. It cannot be perceptibly distorted in shape without breaking, which easily happens. The shell must also be delicate enough to break under the efforts of a new-born chick. However, such smoothly curving shapes are often the result of being liquid rather than solid.

Certainly, a bird's egg was not turned on a lathe like many man-made solids of revolution (such as a chess pawn) and, if it had been turned in a malleable state like clay on a potter's wheel, we would still have the problem of explaining its rigidity. A bird's egg is not fired in a kiln, nor is it hardened by cooling from a molten temperature, as in glass blowing. The shape is characteristic of a flexible surface stretched as a result of the internal pressure of the fluid contents, like a dew drop or an inflated balloon.

Such smooth and simple shapes are known to mathematicians as *minimal surfaces*. A balance is struck between the surface tension (which tries to shrink the shape to a minimum) and the internal pressure of a fixed volume of contents. A perfect sphere is the least surface area enclosing a given volume, but gravity (e.g. in a dew drop), centrifugal spin (e.g. in a planet) or some other force usually enters the picture to distort the surface from a true sphere. Such shapes can only arise in elastic materials. The rigidity of an eggshell is due to a type of electrochemical casting process, which deposits the calcareous material on the flexible and already egg-shaped outer membrane.

EUCLIDEAN EGGS

With these thoughts in mind we return to our task of constructing some idealised plane ovals. We shall call them Euclidean eggs because they will have less to do with birds and science than with Euclid who, as we shall see in the next chapter, was particularly concerned with compass constructions. In a compass construction, the actual scale of the drawing is unimportant, but the relative sizes of the constituent lengths, what we

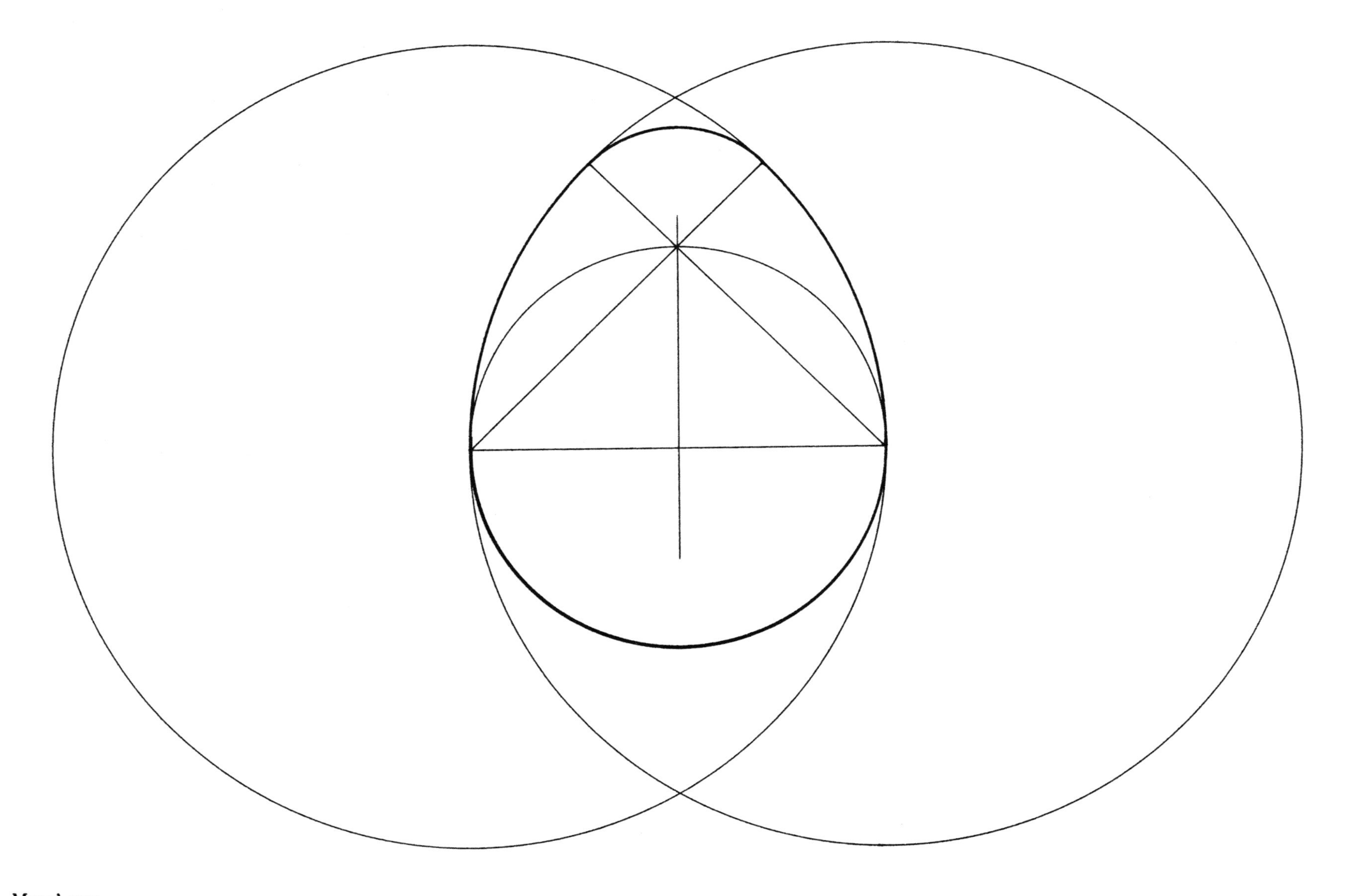

1 Moss' egg

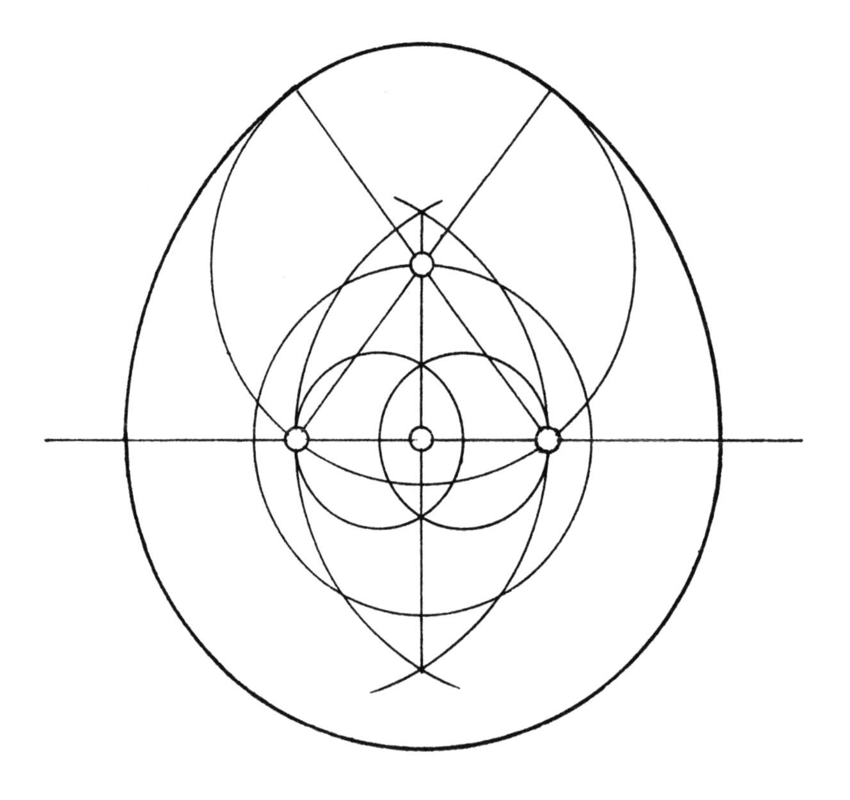

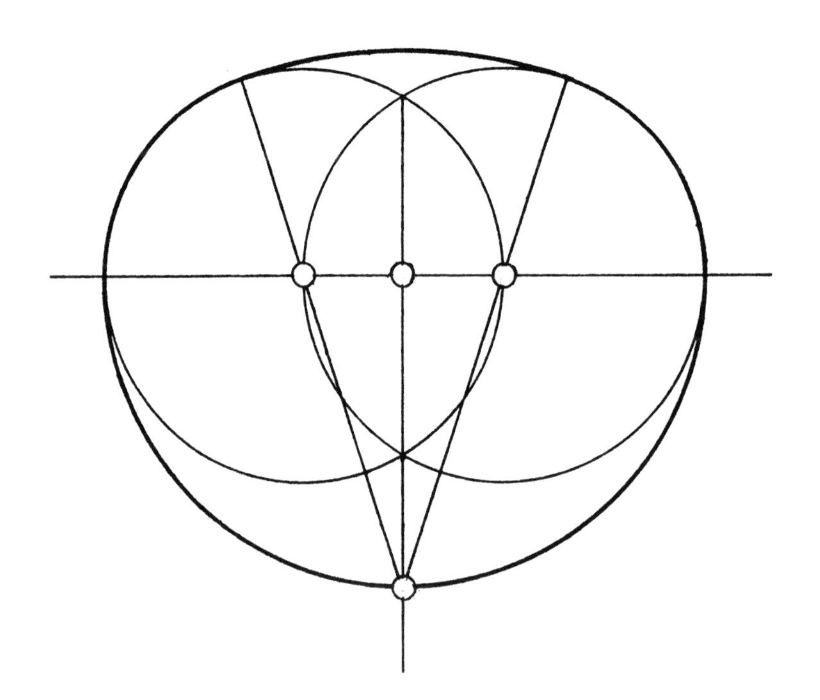

2 Thom's ovals or eggs

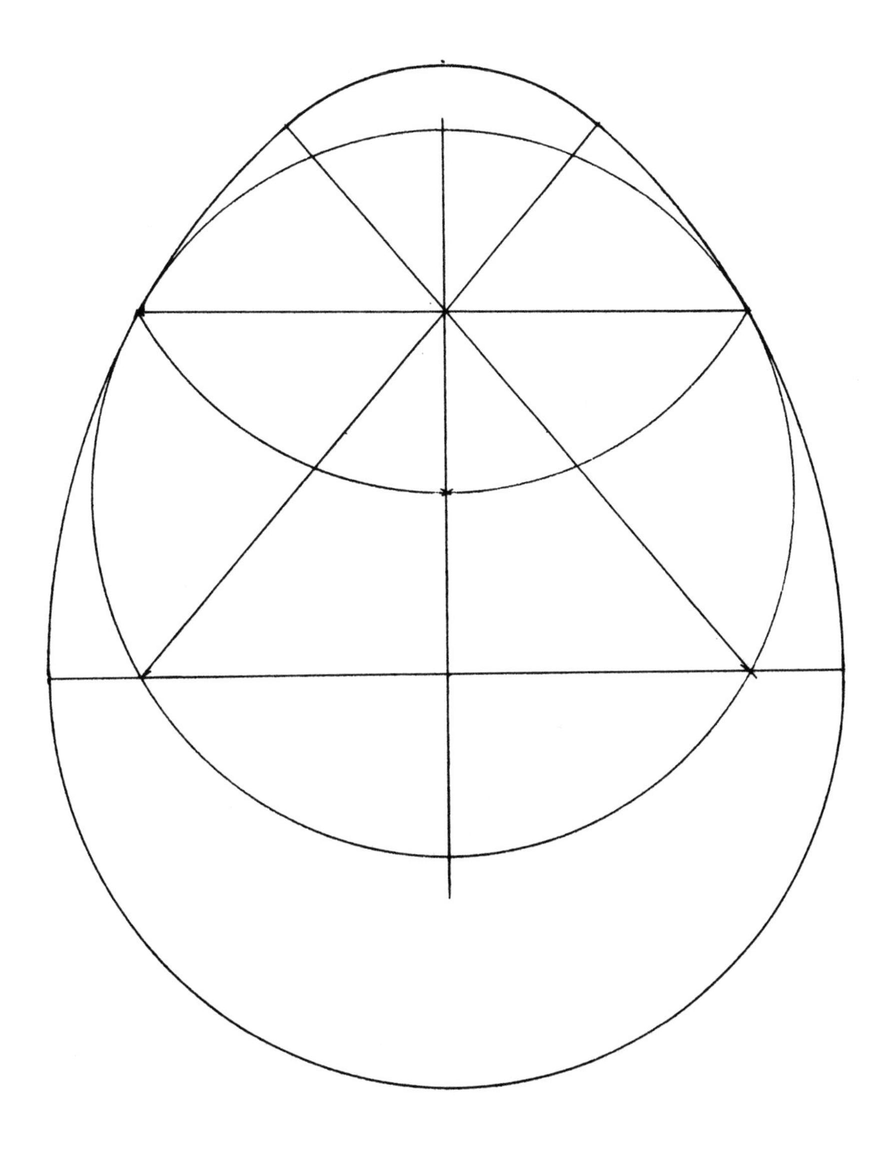

3 $\sqrt{3}:2:\sqrt{7}$

4 Golden egg

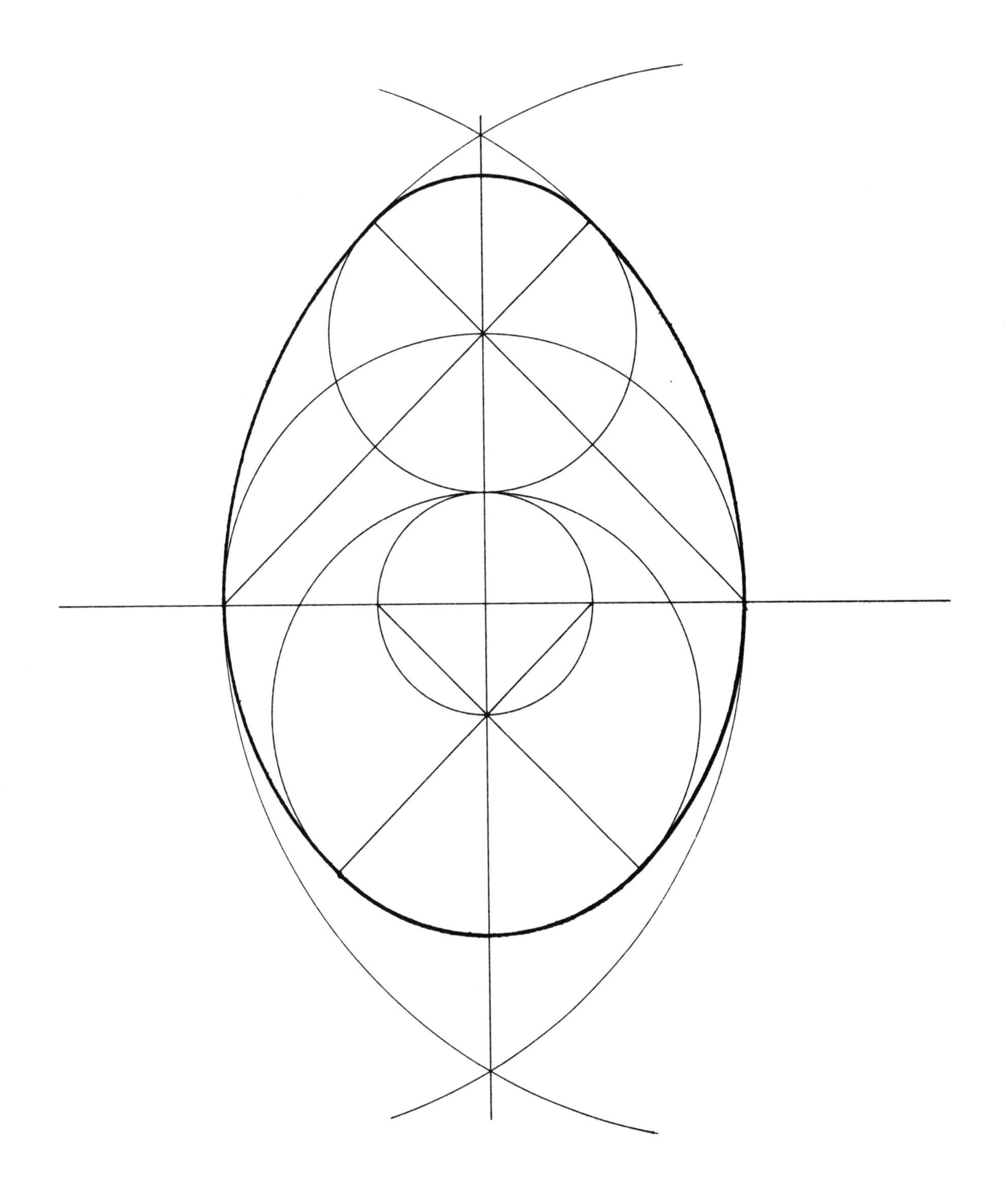

5 Four-point egg

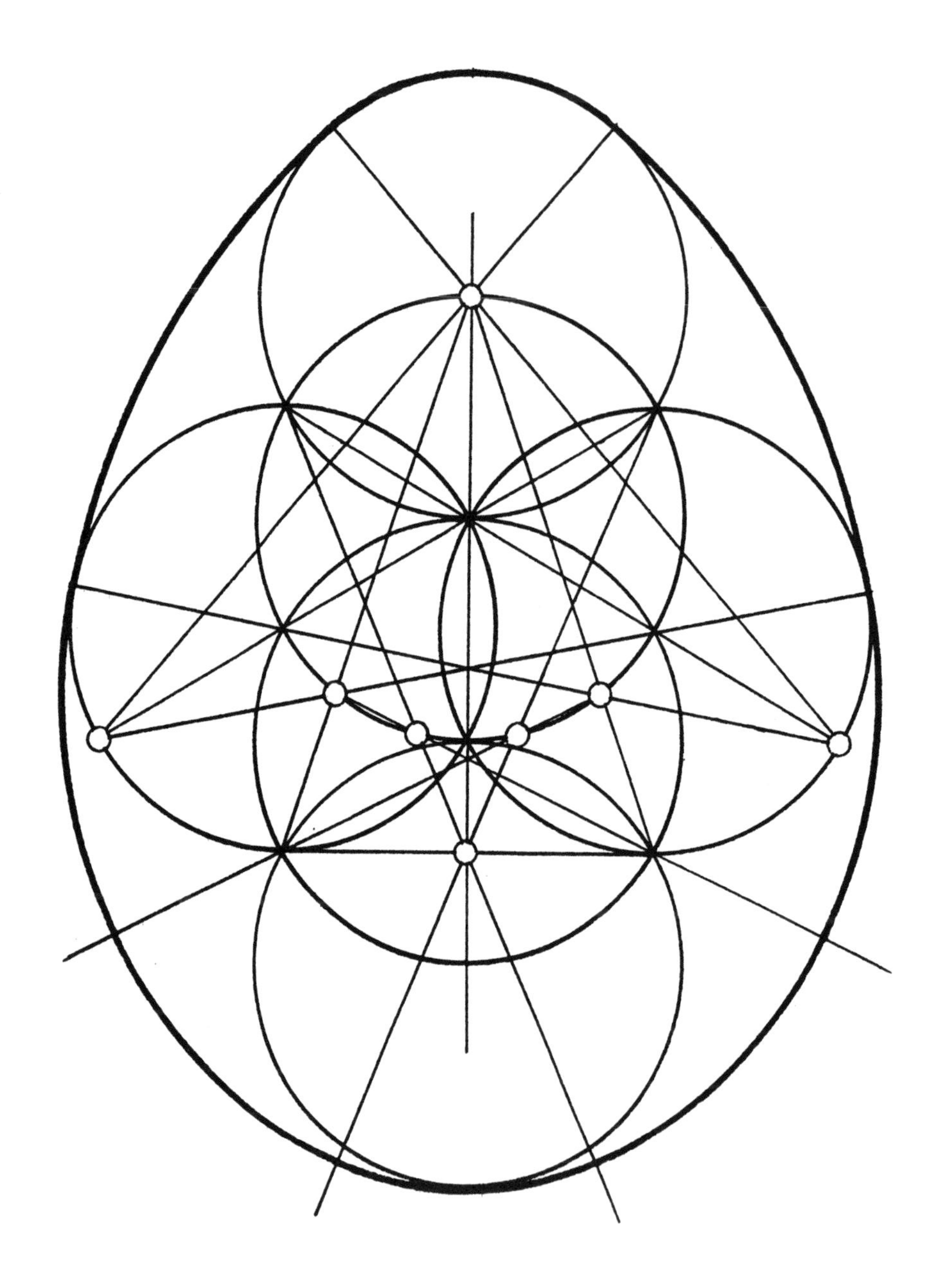

6 Five-point egg

call proportions, are important. A constructable ratio, angle or shape is one which can be repeated exactly by anyone. The relative locations of the compass points must be exactly determined by a recipe for making arcs, lines and intersections.

Drawings **1–6** show a variety of construction designs, ranging from the very simple to the somewhat intricate, starting with some three-arc curves and eventually reaching a five-arc curve. You may like to do several of these constructions, if not all, and then to explore the scope for further possibilities, both egg shapes and other curves.

As we are not concerned with trigonometry in this chapter, questions about the numerical values of the various lengths and angles involved in these constructions are not considered here. If you are interested, you will find them as part of Exercise 9. The subject of egg curves turns up again in Chapter 2, when we move from compass drawing to string drawing and, in particular, from a five-centred egg to an infinite-centred egg.

Exercise 1

Yin yang symbol

1 Learn how to construct each of the egg shapes shown in this chapter.

2 By trial and error and a little imagination, see if you can produce other egg constructions than those given here.

3 Repeat question **2**, but take curves such as arches, letters of the alphabet or chess pieces as the subject rather than eggs.

4 Construct the yin yang symbol.

5 Find a good example of a well-defined curve and see whether you can superimpose a number of circles on various parts of it.

6 The egg in Figure 1.1C is called Cundy and Rollett's egg, as it appeared in their book of mathematical models. If it is made in card or plastic sheet in separate parts it forms a dissection puzzle. Not only do you have the problem of putting it together, which is fairly easy, but also the challenge of making various figures, as shown. Work out how to construct this egg shape.

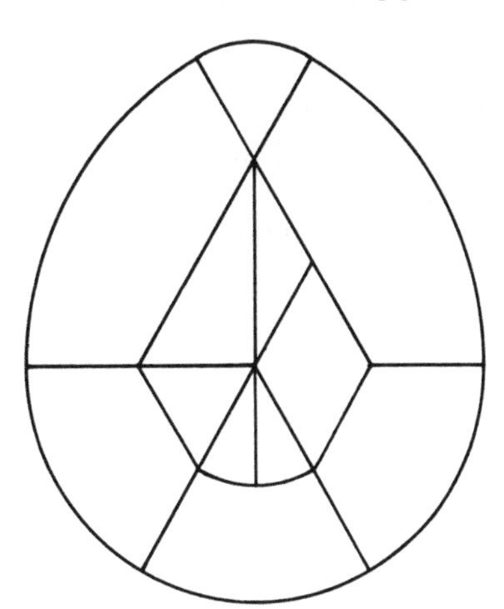

Figure 1.1C

Reading

Critchlow, 1979; Cundy and Rollett, 1961; Dixon, 1982; Pedoe, 1976; Thom, 1967.

1.2 Euclidean constructions

THE DRAWINGS

At the heart of Euclid's geometry of line and circle, there is the following question: given only compass and rule, what is it possible to construct in terms of length and angle, i.e. what parts of a given length and what parts of a circle? For example, is it possible to construct one-tenth of a given line and one-tenth of a given circle? The answer is yes in both cases. However, what about one-ninth of a given line or a given circle? The answers this time are that one-ninth of a given line can be constructed but that one-ninth of a given circle cannot.

In this section the following key constructions with which Euclid answers the general question above (as well as others) are presented.

- **a** To divide a circle into six (and three) equal parts (Drawing **7**).
- **b** To divide a circle into four equal parts (Drawing **8**).
- **c** To divide a circle into five and ten equal parts (Drawings **9** and **10** respectively).
- **d** To bisect any given angle (Drawing **11**).
- **e** To construct a right angle in a semicircle (Drawing **12**).
- **f** To construct the tangent from a given point to a given circle (Drawing **13**).
- **g** To bisect any given line (Drawing **14**).
- **h** To divide a line into any number of equal parts.
- **i** To divide a line into any given ratio of parts (Drawing **15**).
- **j** To construct the square root of any given length (Drawing **16**).
- **k** To construct the golden ratio (Drawing **17**).

All these constructions are easy to perform and most are easy to understand. Constructions **a**, **b**, **d** and **g** are the easiest to grasp because they involve the visually clear principles of congruence (identical sizes) and symmetry. Constructions **h** and **i** involve the principle of similarity (identical proportions) which is also visually clear. However, the constructions involving square roots (which include the golden ratio and the regular pentagon) rely for their explanation on the well-known theorem of Pythagoras. This principle is not visually obvious and it will be described more fully in Section 4.1.

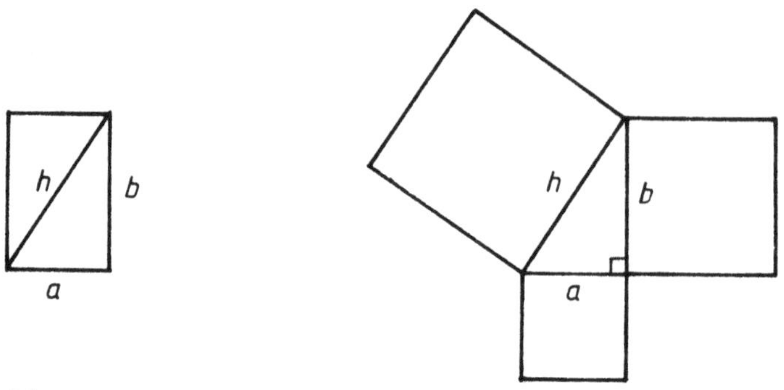

Figure 1.2A

The theorem of Pythagoras (Figure 1.2A) is as follows:

$$h^2 = a^2 + b^2$$

or

$$h = \sqrt{(a^2 + b^2)}$$

where a, b and h are the sides of a right-angled triangle, as shown.

CONSTRUCTION SHORT-CUTS

Euclid says that in order to draw a specific line you need two points and, similarly, that to open your compass to a specific radius you need two points. Nevertheless, you will find that it is easy, and almost as natural, to perform any of the following constructions.

- **a** To draw a tangent through a given point to a circle (Figure 1.2B).
- **b** To draw a tangent common to two given circles (Figure 1.2B).
- **c** To draw a circle about a given centre to touch a given line (Figure 1.2B). (This is somewhat trickier, as it involves opening and closing the compass while gently swaying it backwards and forwards until a good fit is obtained.)

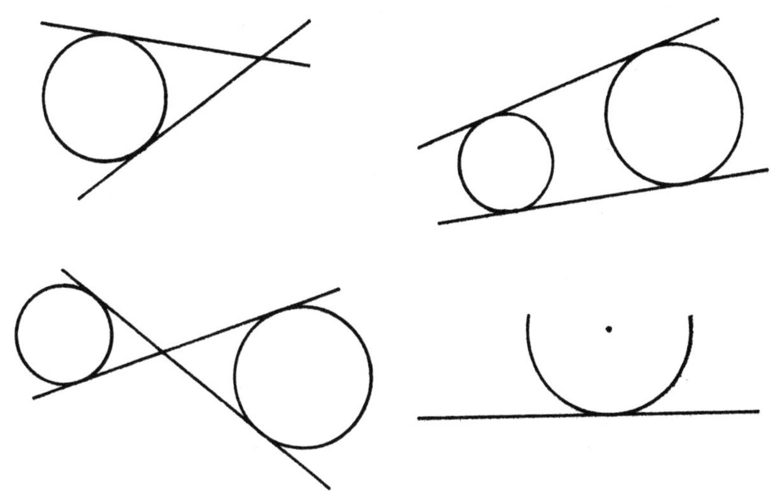

Figure 1.2B

- **d** To draw a line through a given point parallel to a given line (Figure 1.2C). (This can be done very quickly by combining **a** and **c**.)

Figure 1.2C

You may like to use these short-cuts as they greatly simplify many of the constructions but, *be warned*, these short-cuts provide only the lines and circle in each case; they do not locate the point of tangency. The problems of finding points of tangency will be given in Exercise 2.

With a good compass and skilful handling, it is also possible to find the mid-point of a given line segment and to draw the circle. This is done by trial and error involving both the location of the mid-point and the length of radius. First, place the compass point at a likely position, and open the compass to reach from there to one end of the line segment. Next, without removing the compass point, swing the compass round to see whether it reaches the other end of the line segment. It will probably miss by a certain amount. Now re-adjust the compass radius to reduce the error by half. Finally move the compass point by the other half of the error and repeat the procedure. You should be able to find the centre in three or four repeats of the procedure as the error is rapidly reduced in a systematic matter. But it does require a delicate touch so that the trial centres do not result in the paper being fouled by the compass point.

CONSTRUCTABLE LENGTHS

In a Euclidean construction it is as if the inch, the centimetre and all other standard units of length had never been invented, for they play no part. We start each drawing by opening the compass to an *arbitrary* radius which is taken to represent the unit of length for that drawing. We are not concerned with the actual size of a drawing, but only with the relative lengths (proportions) involved. So, for example, to construct a length of $\sqrt{2}$, we can construct the diagonal of a square whose side length equals 1 unit (Figure 1.2D).

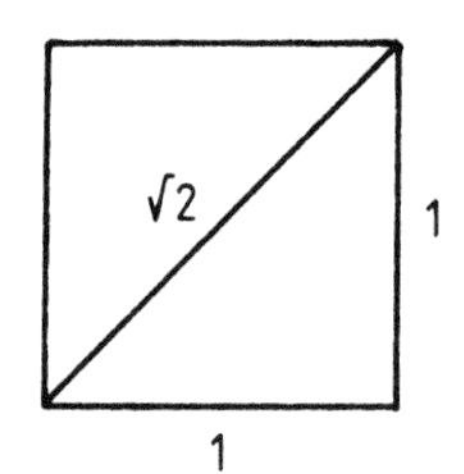

Figure 1.2D

Let us see the range of numbers which can be constructed.

- **a** We can construct any whole number length by using a compass to 'step off' the unit radius along a line as many times as we please.
- **b** We can bisect any given line to obtain half-lengths, quarter-lengths and so on (Drawing 14).
- **c** We can divide a line into any number of equal parts by parallel projection, to obtain all fractions (Drawing 15).
- **d** The theorem of Pythagoras tells us how to construct the square root of any number, including square roots of square roots and so on (Drawing 16).

This is an impressive list of possibilities. Yet it turns out that there are plenty of numbers which it is impossible to construct exactly, including π and cube roots, for example. (The subtle challenge of constructing very good approximations to such impossible constructions is the subject of Section 1.3.)

CONSTRUCTABLE ANGLES

Whereas lengths are relative, angles are absolute. Angle measures the inclination between two lines, or the amount of turning required to change from facing one direction to facing a second direction. Two lines meeting at a point provide our clearest image of what an angle looks like. If we draw a circle centred on their point of intersection, the angle in question is the part of the circle cut by the two lines (Figure 1.2E). In other words, the absolute unit of angle is one circle: a complete revolution.

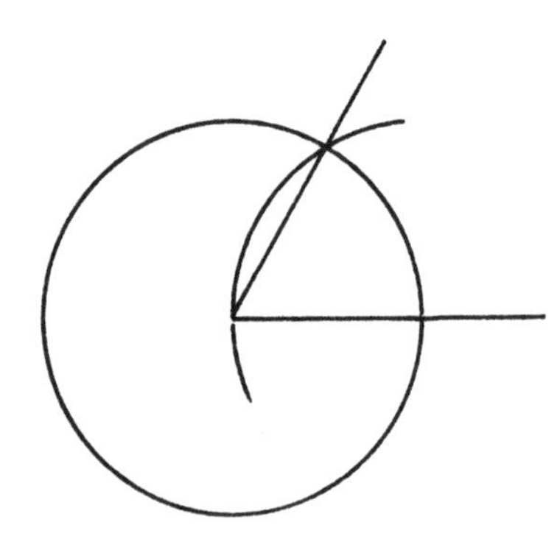

Figure 1.2E

Long before Euclid's time the Babylonians had divided the circle into 360°. It is important for us to realise that their choice of 360 had little or nothing to do with geometry. It was to do with arithmetical convenience: firstly, because it avoids having to talk in terms of fractions of a circle all the time and, secondly, because 360 can be divided exactly by 2, 3, 4, 5, 6, 8, 9, 10, 12, 15, . . . (and what other whole numbers?). This great convenience with arithmetical division explains why most people continue to use this ancient system.

We shall use degrees to name our angles but we shall not use a protractor to measure or construct them. What angles can we construct?

- **a** Starting with a circle of any radius, use a compass to 'step' this radius around the circumference — you should find that it goes exactly six times. (Why is this so?)
- **b** By constructing the perpendicular bisector of a diameter of a circle, we can divide the circle into four equal parts.
- **c** Once shown how, we can construct a regular pentagon in a circle, giving us fifths of a circle.
- **d** It is possible to construct thirtieths of a circle by inscribing a regular hexagon and a regular pentagon in the same circle. (This is because the difference between a fifth and a sixth is a thirtieth.)
- **e** It is an easy matter to bisect any given angle.

All in all, this gives us the following constructable angles:

60°, 30°, 15°, ...
90°, 45°, 22.5°, ...
72°, 36°, 18°, ...
24°, 12°, 6°, ...

Angles which we cannot construct include 1°, 10°, 40°. We cannot construct the regular heptagon (seven sides) or the regular enneagon (also known as a nonagon) (nine sides) and many other regular polygons. The complete list of whole number divisions of a circle which are impossible to construct begins thus: 7, 9, 11, 13, 14, 18, ... It turns out that an exact 17-fold division is possible, but, as this was not discovered for 2000 years after Euclid and the constructions in this section, I have made it the subject of Section 1.4.

Although it is easy to bisect any given angle, it is not possible to trisect any given angle, neither is it possible to divide an angle into any number or ratio of parts other than 2, 4, 8 and so on. (Again, if you are interested in approximating the regular heptagon and enneagon, you will find details in Section 1.3.)

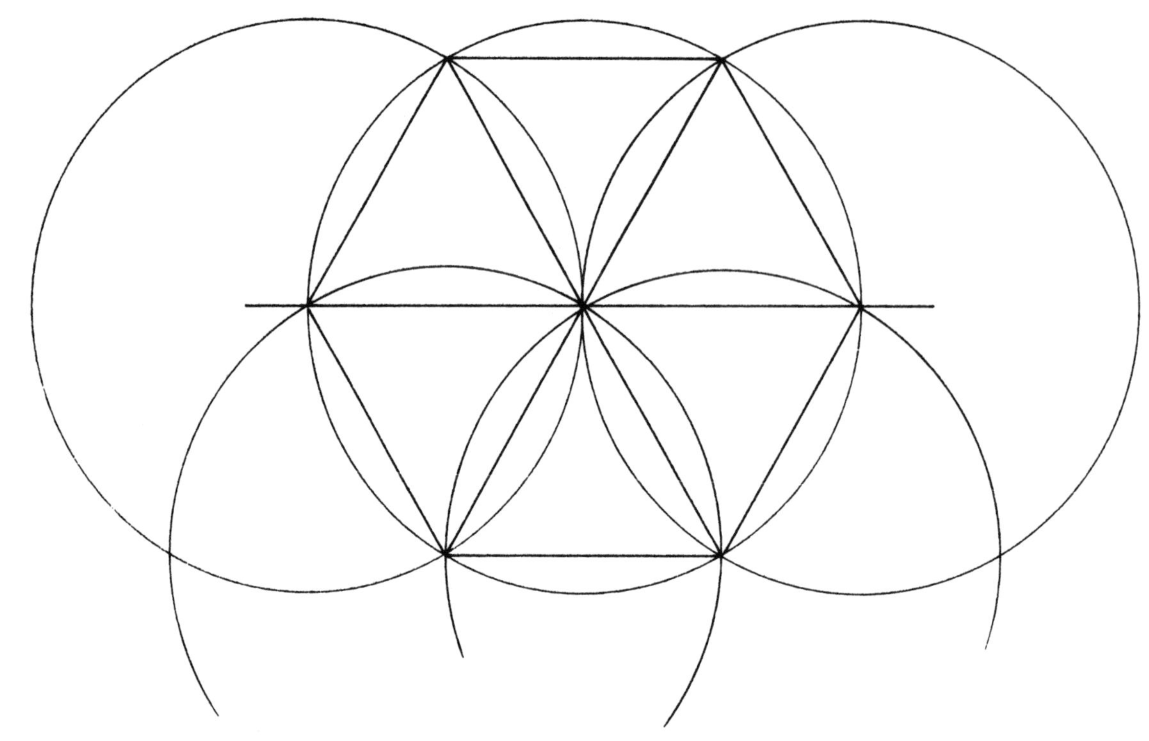

7 Hexagon

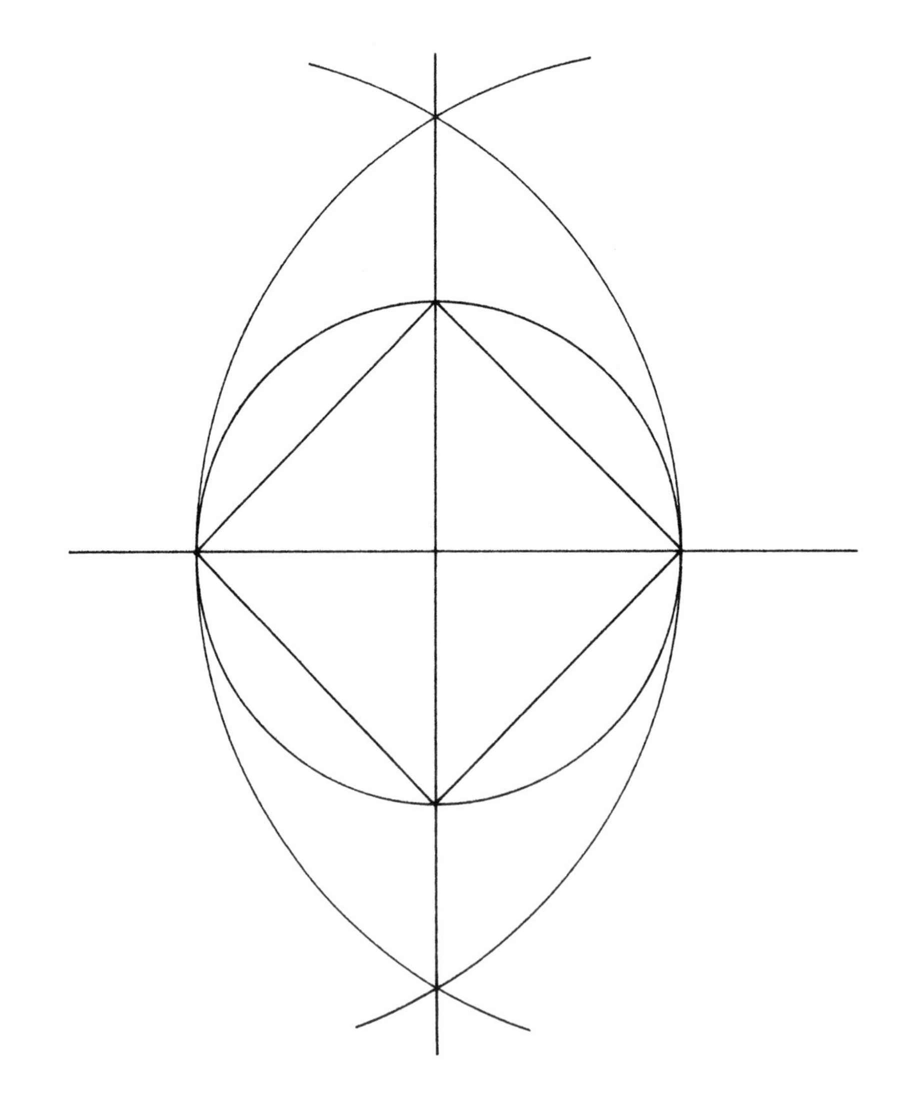

8 Square

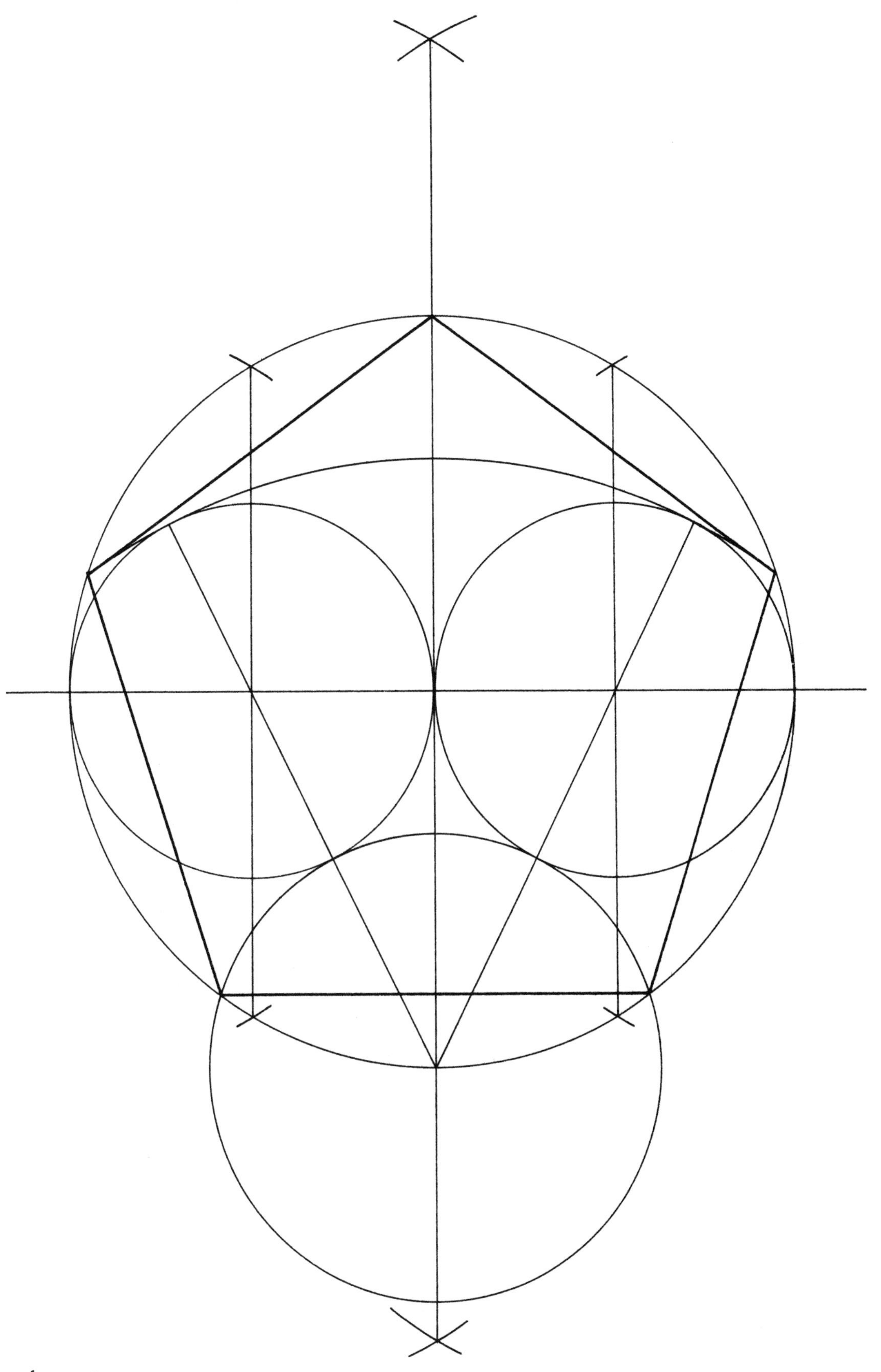

9 Regular pentagon

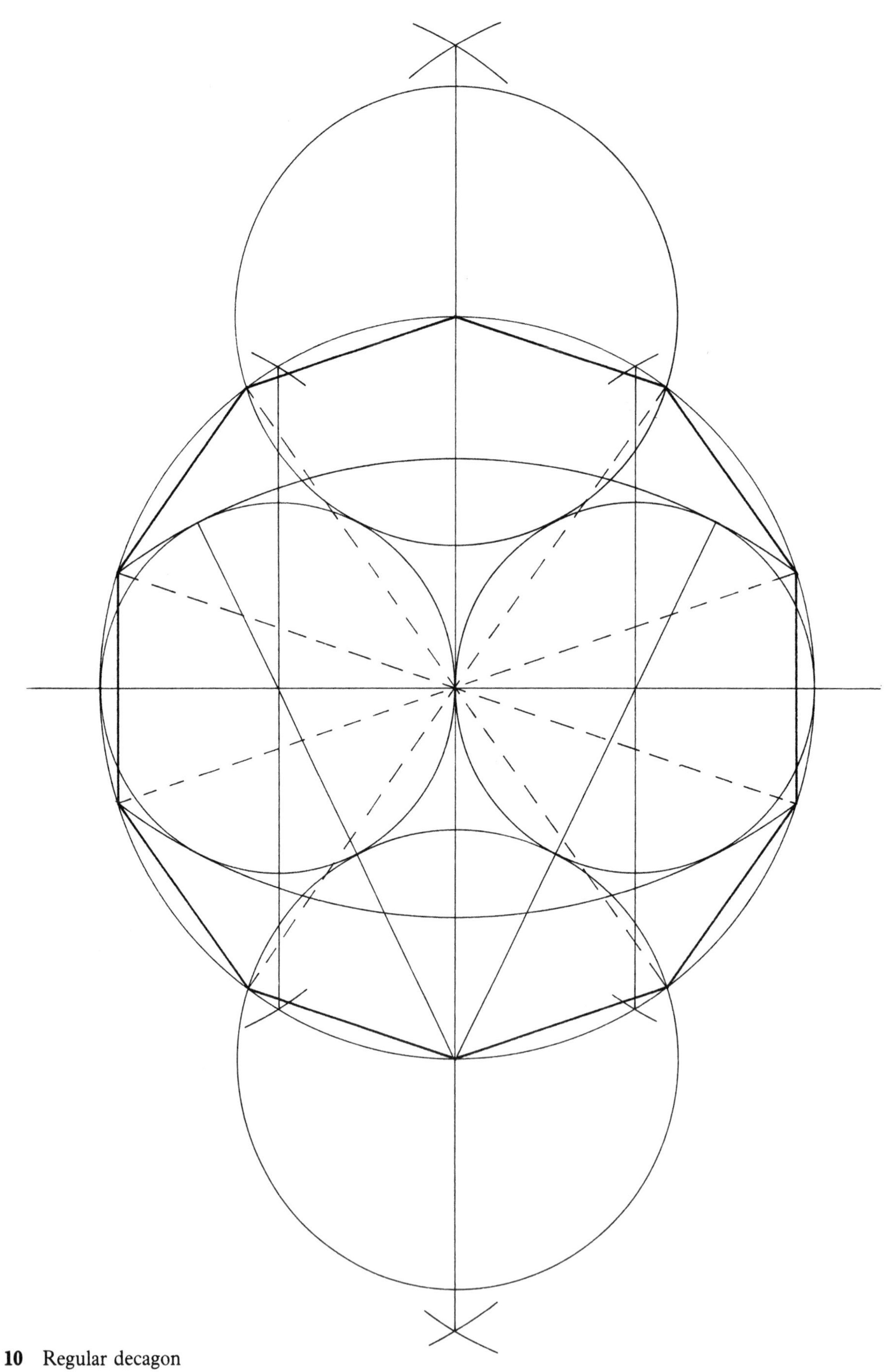

10 Regular decagon

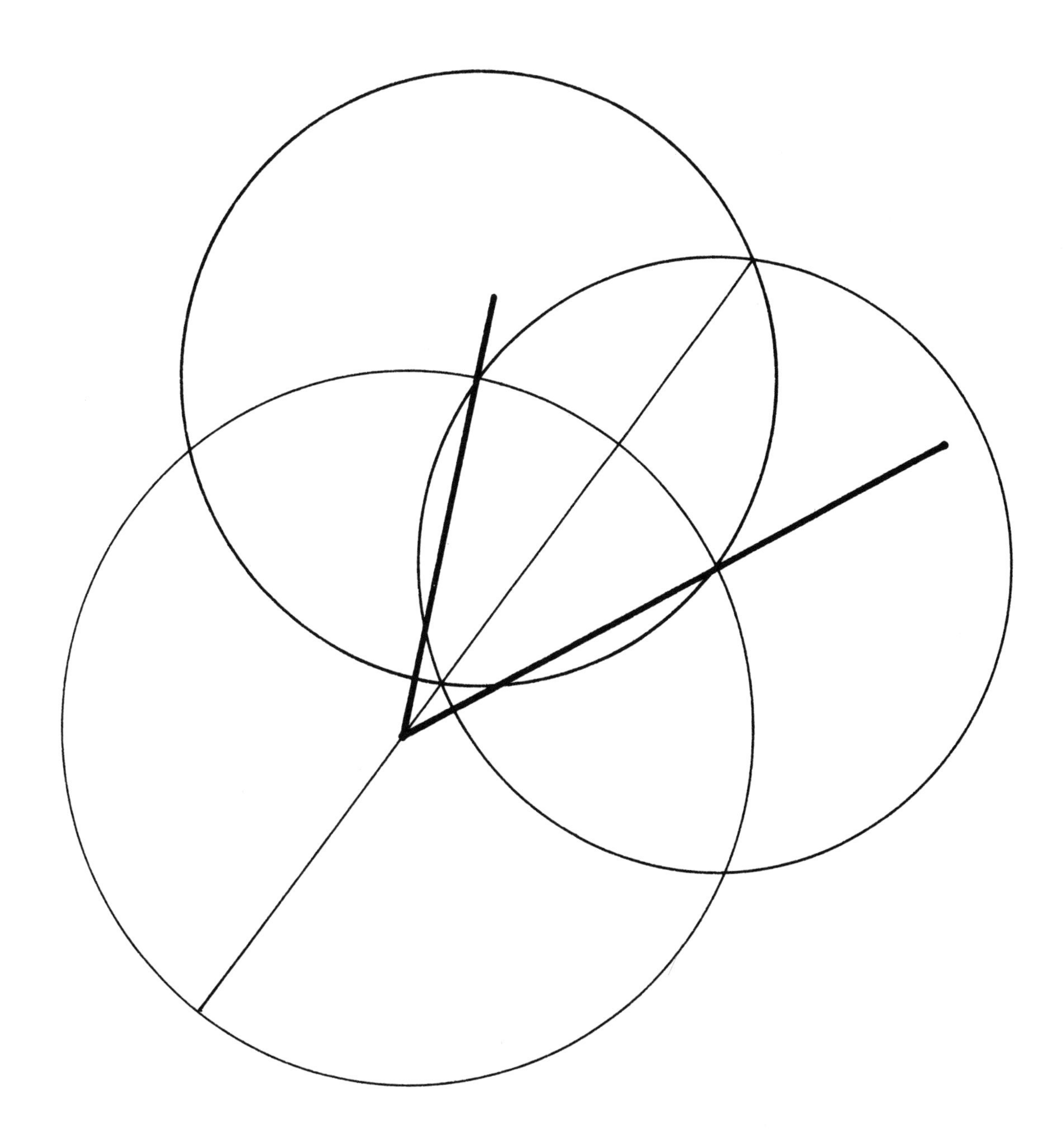

11 Angle bisection

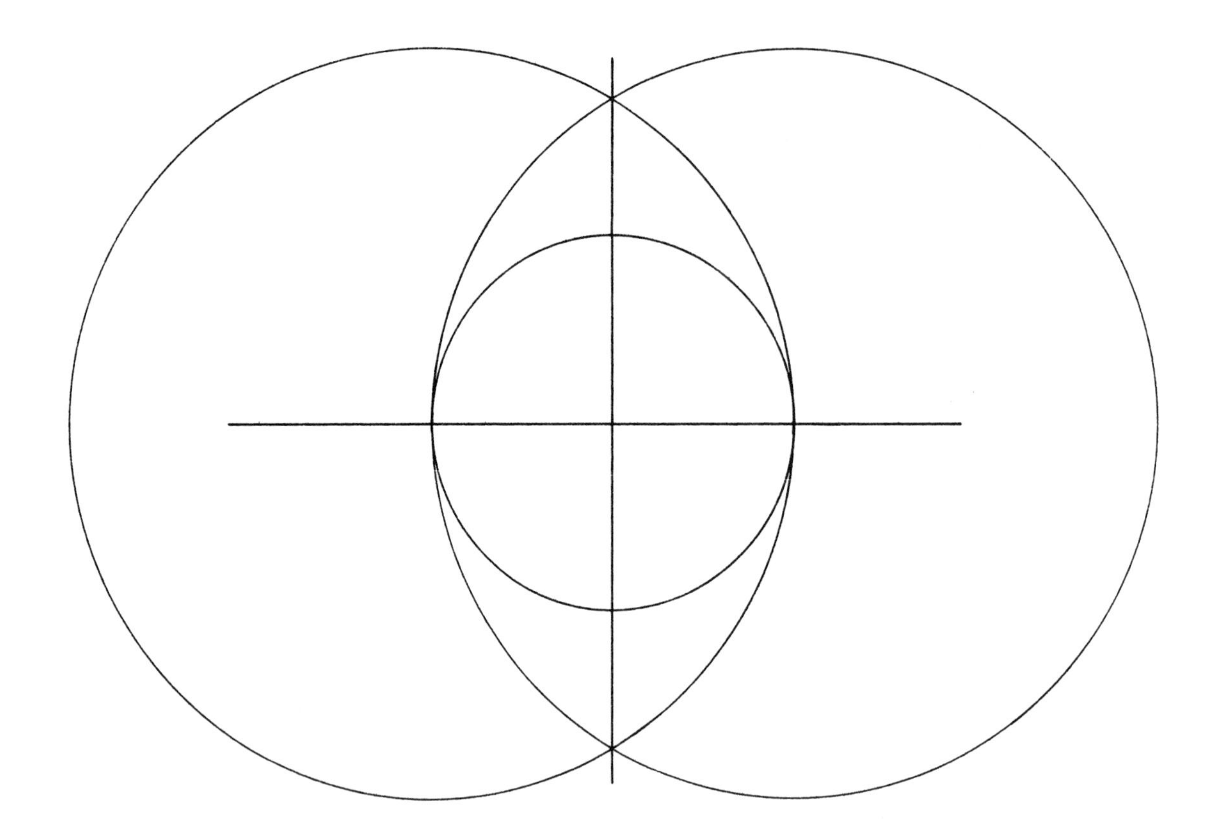

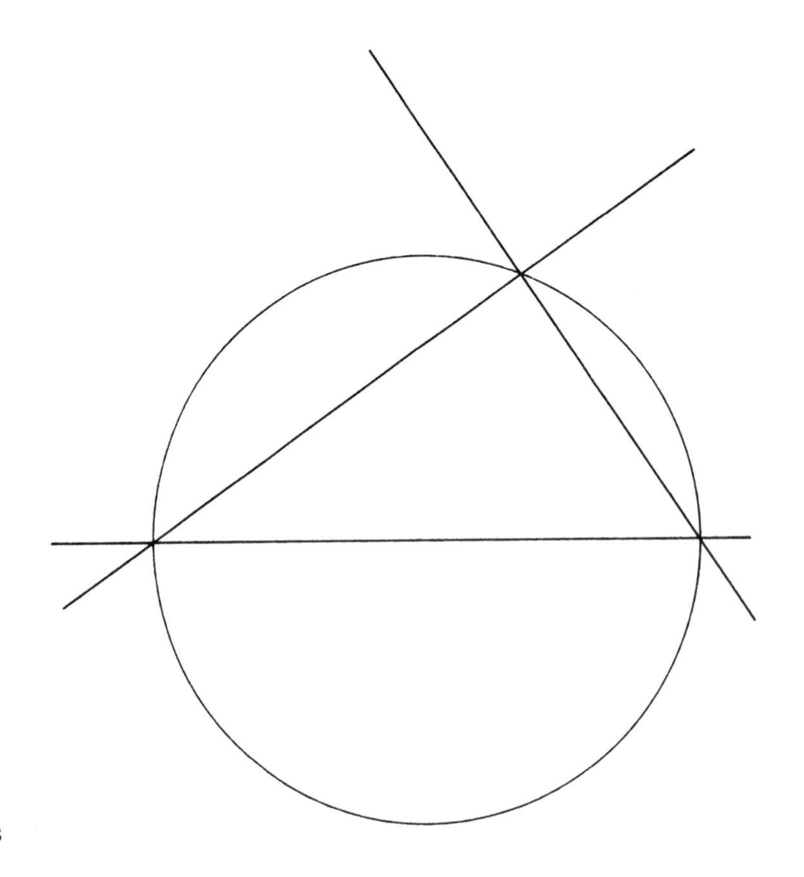

12 Right angles

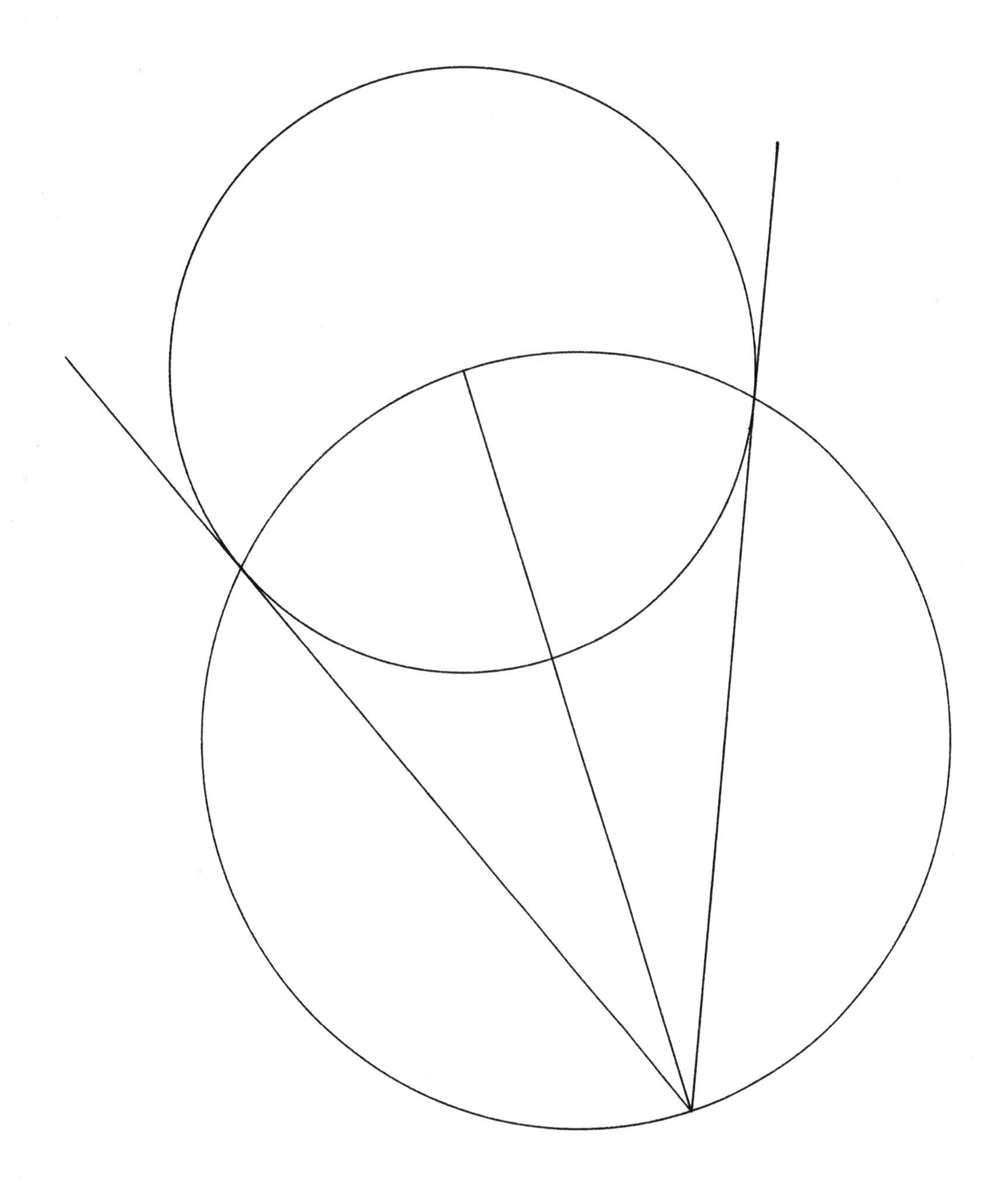

13 Tangents to a circle

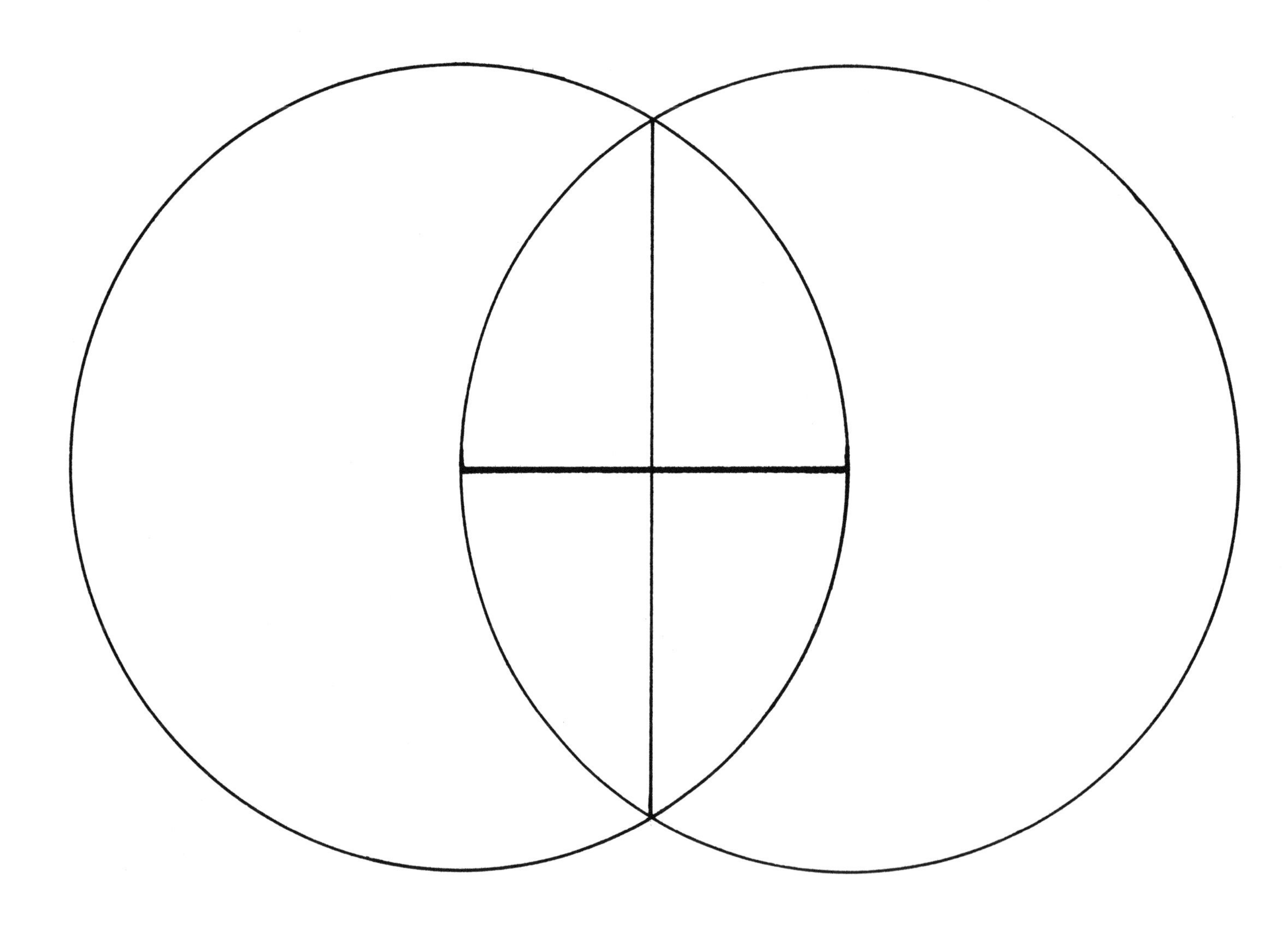

14 Line bisection

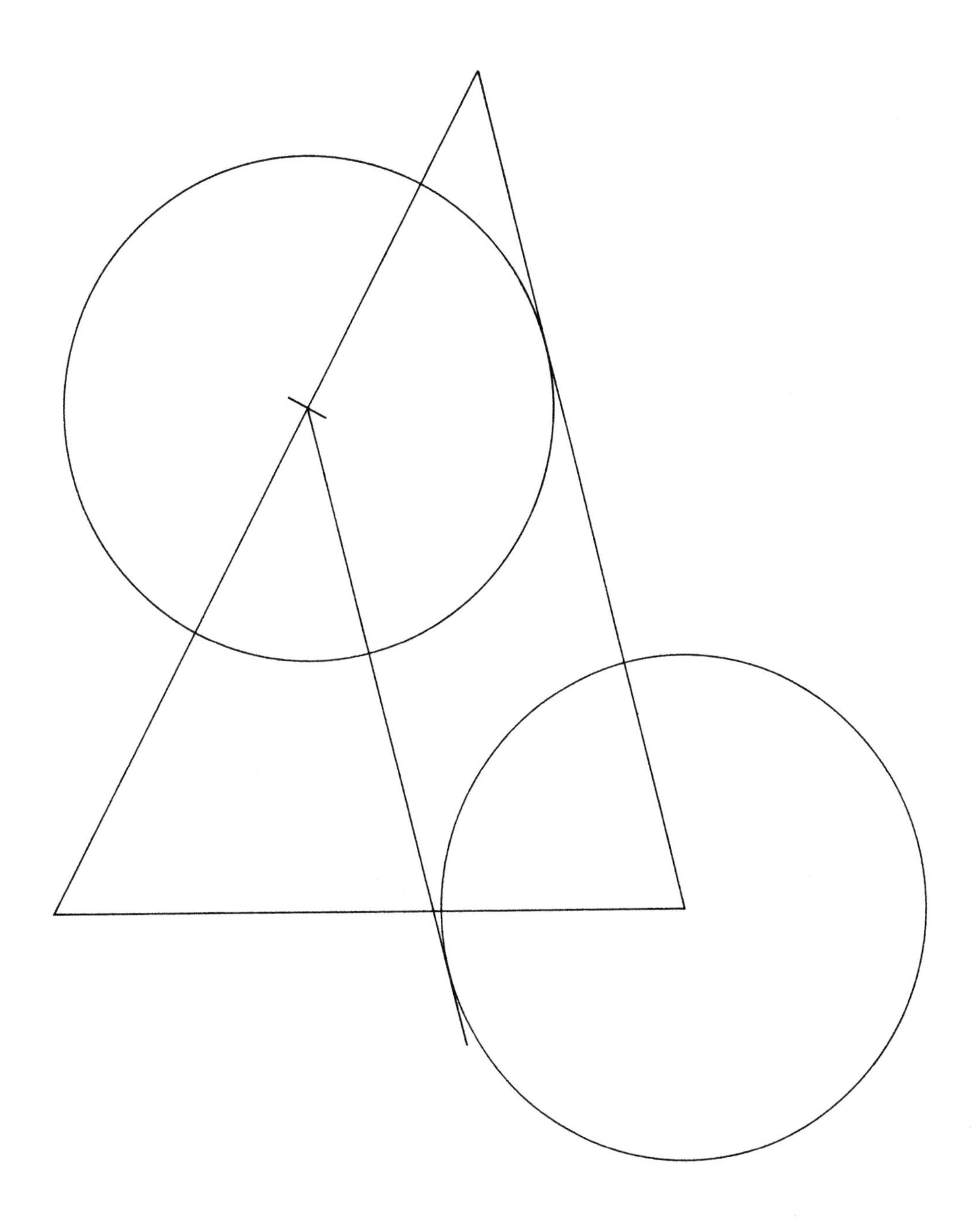

15 Line division by a given ratio

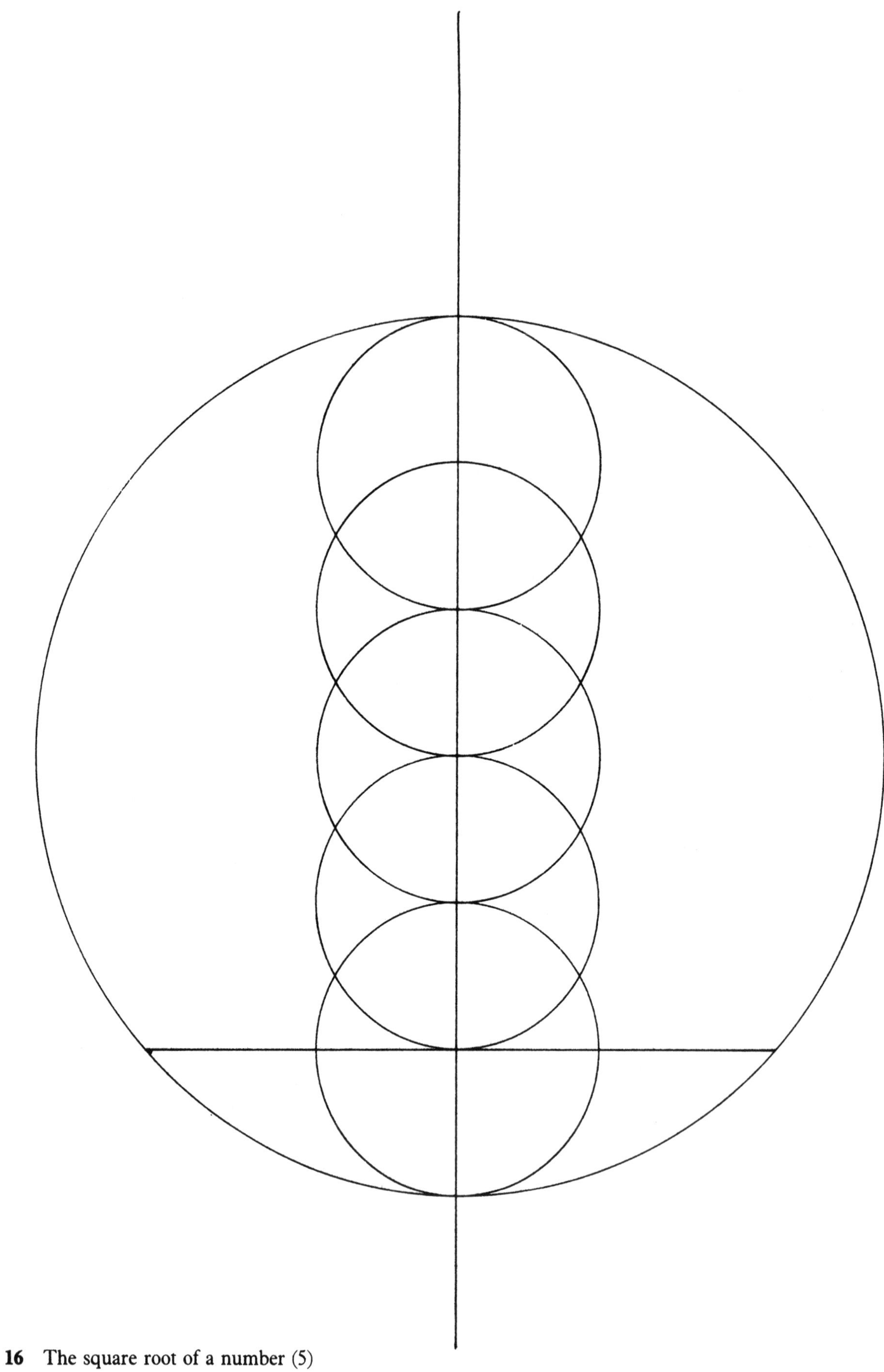

16 The square root of a number (5)

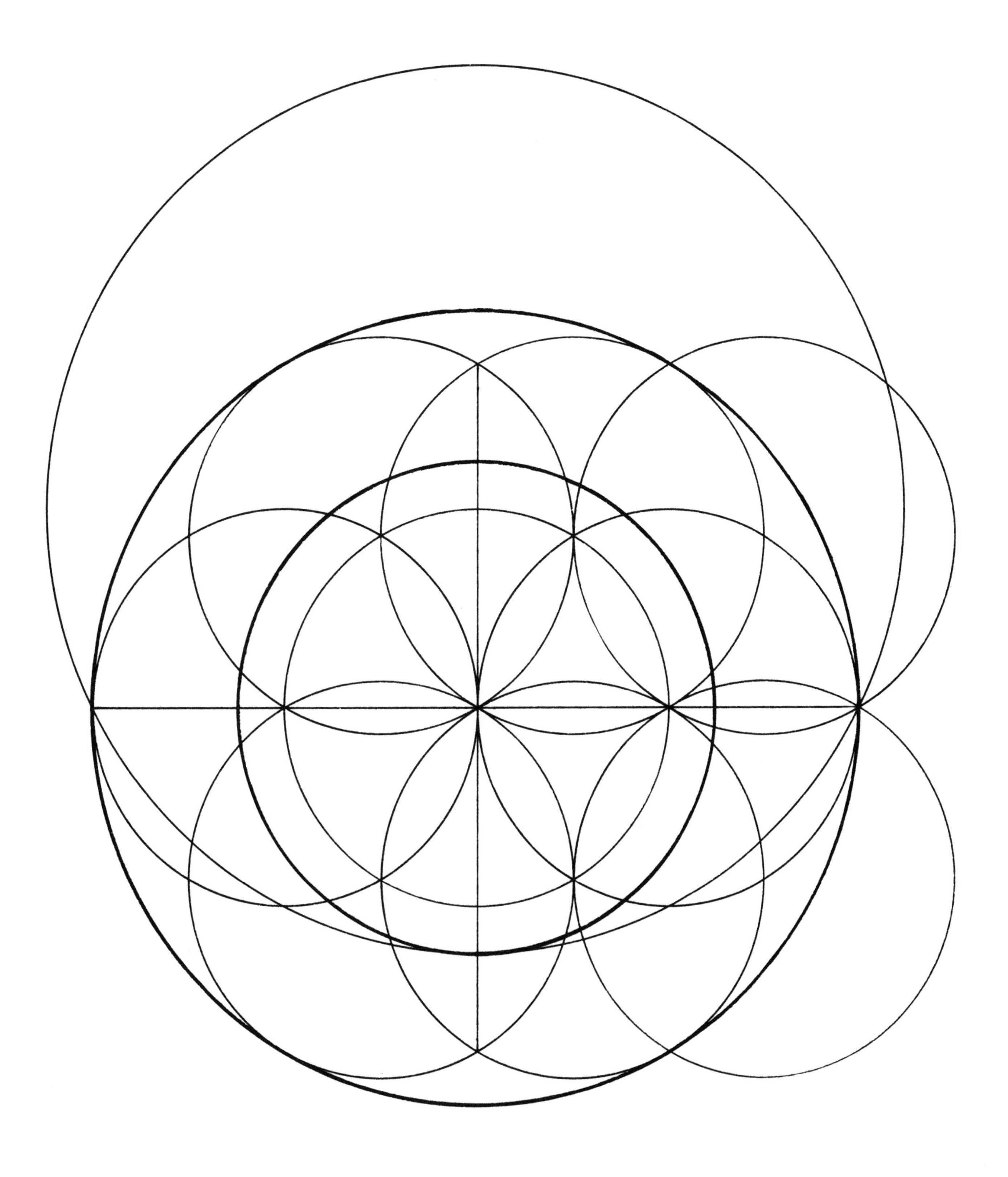

17 Golden ratio

EUCLIDEAN EXPLANATIONS

In this book of mathematical drawings, we are more concerned with doing the drawings and learning the methods than we are with explaining them. It is easier to construct the regular pentagon than it is to understand why the method works. Euclid himself, being a mathematician (in fact the most famous mathematician of all time), was more interested in the explanations than in the drawings. For those readers who are also interested in the explanations, what follows is a brief account of the ideas behind the Euclidean constructions that we have been looking at in this chapter. If you wish to know more about these and other well-known principles of elementary geometry you should study any one of the very many beginner's books on the subject.

All modern books of elementary geometry are based on Euclid's book, *The Elements*. Euclid's fame is due entirely to this book. We know very little else about him, apart from the stories that say he lived and worked in Alexandria when it was the centre of Greek learning, around 300 BC. *The Elements* is by far the oldest book of mathematics that we have. It has been the favourite textbook for beginners in almost every generation since it was first published. Euclid's original book contains 465 theorems and even in translation it seems difficult for us to read. We now have the added advantage of being able to use algebra (which was not invented until long after Euclid's time) to say many things much more briefly and clearly. So, for clarity and simplicity, it is best to read a modern book of elementary geometry, which would typically select about 40 or 50 of the best theorems to study.

Nevertheless, all modern books of elementary geometry must try to echo Euclid's original purpose. They use a handful of first principles of line and circle (ones which everyone would take to be obvious and unquestionable) to prove all known principles of line and circle (many of which are both surprising and powerful) in a step-by-step logical manner. So we shall start our explanations where Euclid started — with his five axioms of plane geometry.

The first three axioms state what it is possible to draw — they seem very straightforward.

Axiom 1 Between any two points a line can be drawn.
Axiom 2 Any line can be extended in either direction indefinitely.
Axiom 3 A circle of any radius can be drawn around any point as centre.

The next axiom is more mysterious — what it is trying to tell us is that the measure of angle is the same everywhere. This is exactly as we experience things in space, and it would be difficult to imagine them otherwise.

Axiom 4 All right angles are equal.

Perpendicular lines make four right angles where they meet (Figure 1.2F). Of all possible angles, there is something very special about 90°.

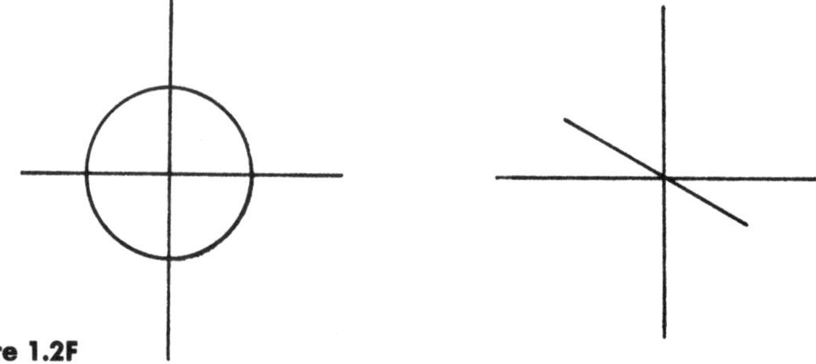

Figure 1.2F

The fall of gravity is perpendicular to the horizontal surface of the earth. Up–down, north–south and east–west are the natural choice for three mutually perpendicular axes for the coordination of three-dimensional space. Algebra and coordinates were invented long after Euclid's time but, although Euclid's geometry stays mainly in the plane, it is clearly intended to include three-dimensional geometry. In contrast, our most frequent means of illustrating three dimensions is in the two-dimensional plane.

A single line makes an angle of 180° at any point along its length (Figure 1.2G). This thought leads directly to the following simple but powerful rule governing angles made where one line meets or cuts another:

$$\alpha + \beta = 180°$$

180° plays a key role in Euclid's fifth and final axiom, which is about parallel and non-parallel lines. How do you show that two lines are travelling in the same direction (parallel) in a given plane? Euclid's answer was to draw a third line to cross them both, called a *transversal*, and to give the following rules.

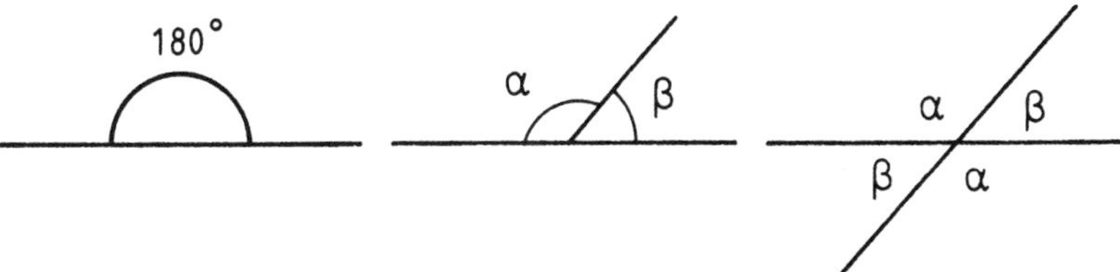

Figure 1.2G

Axiom 5 A pair of lines inclined towards each other will meet somewhere in that direction. Parallel lines keep a constant distance apart (Figure 1.2H). The angles made with a transversal in each case are as follows.

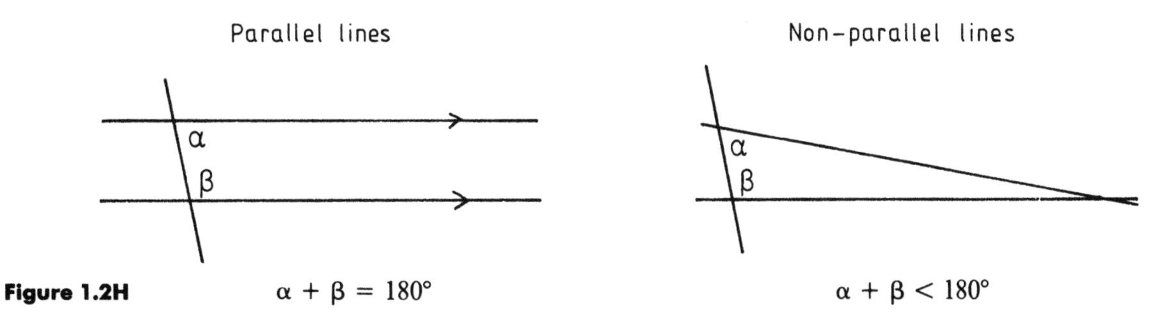

Figure 1.2H

THE ANGLE SUM OF A TRIANGLE

To show the important principle that the sum of the angles of any triangle is 180°, Euclid extended one of its sides and constructed a line parallel to another of its sides, as in Figure 1.2I.

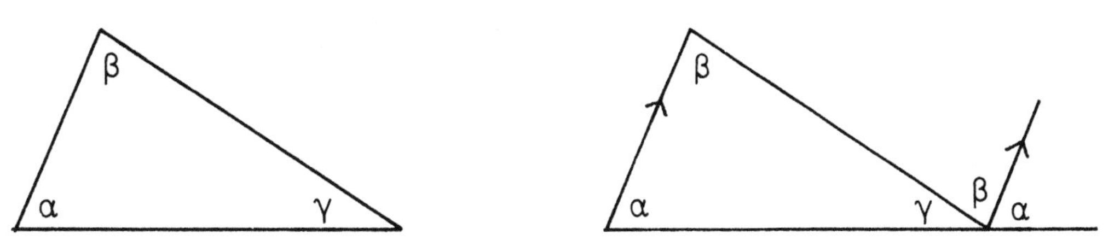

Figure 1.2I

$$\alpha + \beta + \gamma = 180°$$

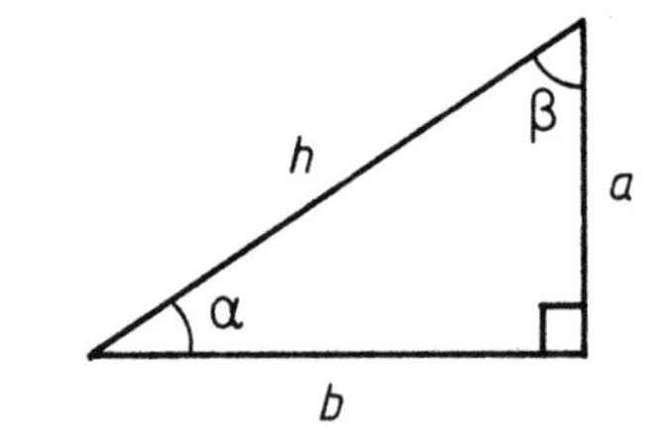

Figure 1.2J

The usefulness of this theorem is that if we know two of the angles of a given triangle we can calculate the third. Right-angled triangles are those with a 90° angle, leaving the other two angles to share 90° between them. They are particularly important because of the rule of Pythagoras and because the rules of sine, cosine and tangent ratios apply to the side lengths (Figure 1.2J).

$$\sin \alpha = \frac{a}{h} \qquad \cos \alpha = \frac{b}{h}$$

Euclid proved (see Section 3.2) and used the theorem of Pythagoras, but the sine and cosine tables (see Section 3.2) play no part in compass geometry.

An *isosceles* triangle is one in which two of the angles are equal, giving two sides equal, and vice versa (Figure 1.2K). The perpendicular bisector of its base divides an isosceles triangle into two equal back-to-back right-angled triangles.

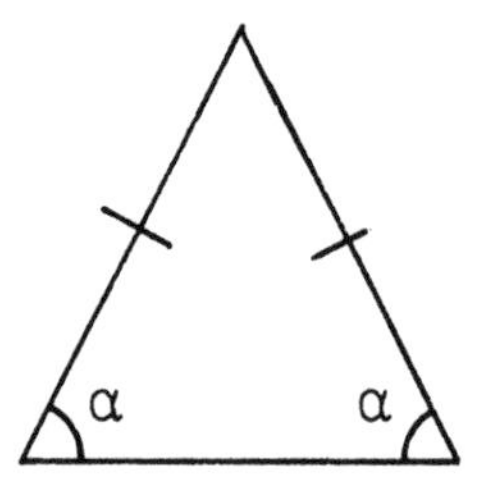

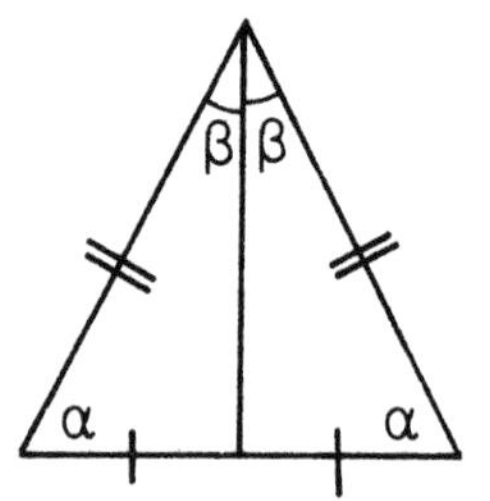

Figure 1.2K

An *equilateral* triangle (Drawing **18** and Figure 1.2L) is one in which all the angles equal 60°, giving three identical sides, and vice versa. Six equilateral triangles make a regular hexagon.

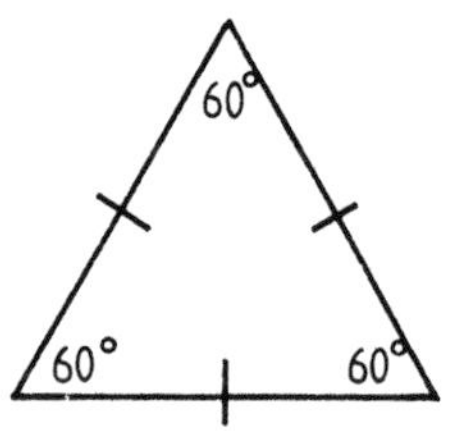

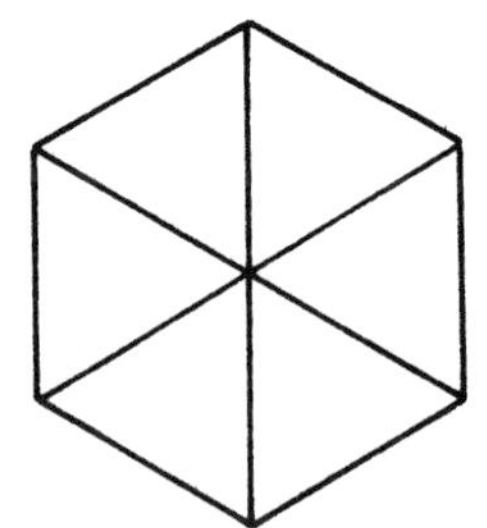

Figure 1.2L

CHORDS AND TANGENTS OF A CIRCLE

At any point on a circle the circumference is perpendicular to the radius. A line drawn through a point on the circumference of a circle perpendicular to the radius at that point is a *tangent* to the circle (Figure 1.2M). We say that a tangent *meets* the circle because only one point of the line is on the circle, while all the others are outside the circle. Any other line through a point on the circle must *cut* it twice, making a *chord* (Figure 1.2M). A chord divides a circle into two segments; in the special case when the chord is a diameter, these segments are semicircles.

A chord of a circle together with its two end-point radii make an isosceles triangle (Figure 1.2N). Similarly, a chord of a circle and its two end-point tangents make an isosceles triangle. The last of the diagrams in Figure 1.2N includes the perpendicular bisector; how many right-angled triangles can you find?

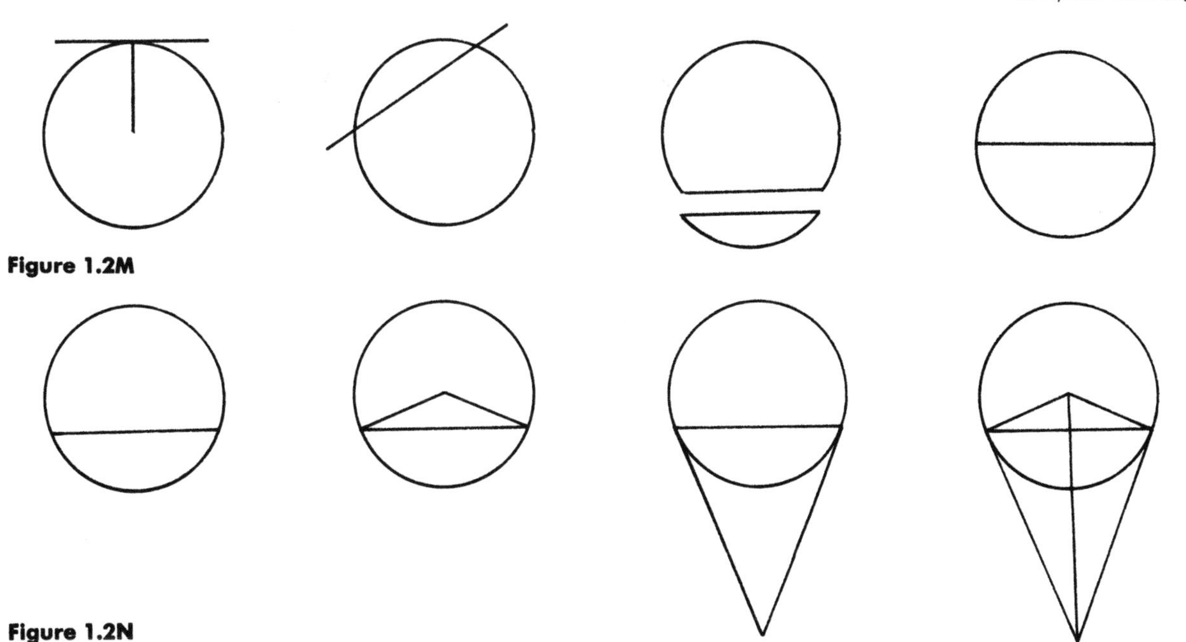

Figure 1.2M

Figure 1.2N

THE ANGLE IN A SEMICIRCLE

It is quite easy to show that the angle in a semicircle (as shown in Figure 1.2O) is a right angle. By drawing a radius to the angle in question, you create two isosceles triangles:

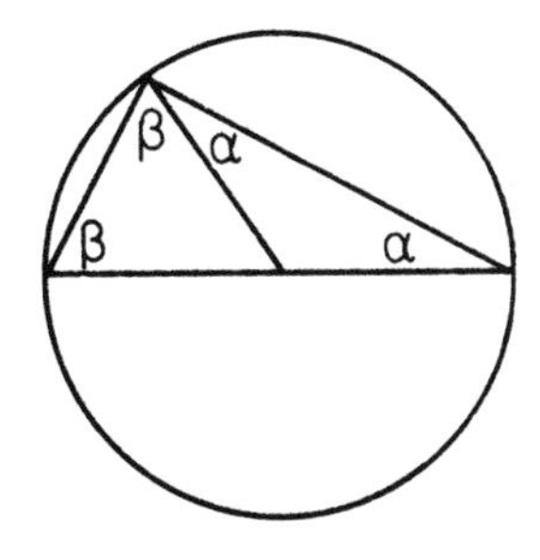

Figure 1.2O

The sum of the angles of the large triangle.

$$\beta + \beta + \alpha + \alpha = 180°$$

Therefore $\alpha + \beta = 90°$

Having established that the angle in a semicircle is a right angle, we can go on to prove by similar triangles that the perpendicular from the right angle to the diameter of the circle obeys the following rule (Figure 1.2P):

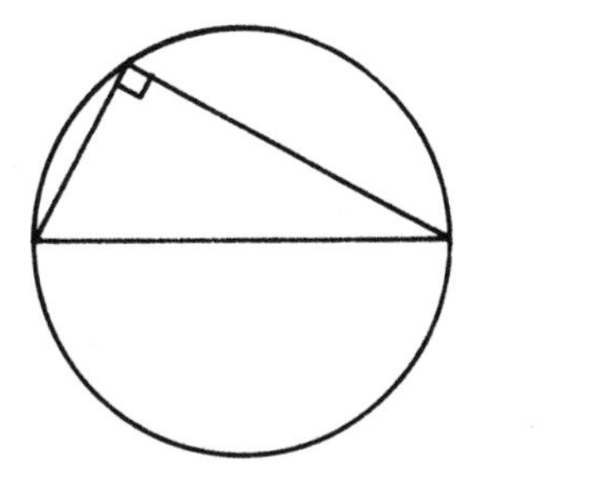

Figure 1.2P

$$\frac{a}{x} = \frac{x}{b}$$

Therefore $x^2 = ab$

It is this principle which allows us to construct the square root of any length L: by putting $a = 1$ and $b = L$ in the above equation it follows that $x = \sqrt{L}$.

THE GOLDEN RATIO
$\tau = \frac{1}{2} + \frac{1}{2}\sqrt{5}$
$= 1.6180\,034\ldots$

The golden ratio τ (tau) is one of those numbers which, like π and e, keeps turning up in every corner of mathematics. It has some surprising roles to play in nature, as we shall see in Section 5.1. The frequent claims that it is the secret of pleasing proportions in art and architecture are mostly exaggerated, but there may be an element of truth in them. Euclid showed how to construct it and also that it is a constituent part of the regular decagon (Figure 1.2Q).

The definition of the golden ratio is best explained in terms of the golden rectangle. A rectangle is golden if a smaller rectangle of the same shape is left when the square which fits the shorter side is removed (Figure 1.2R).

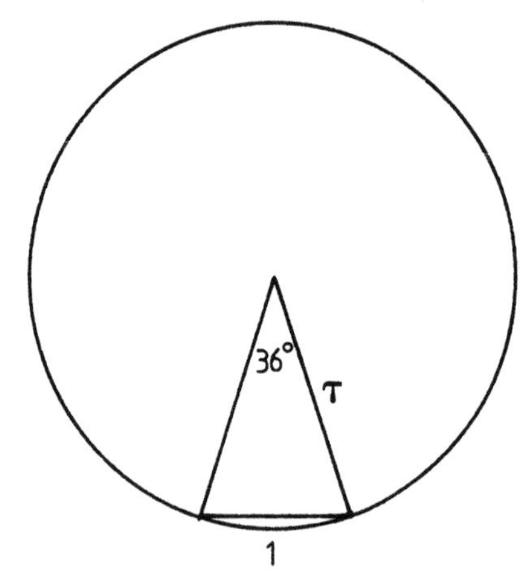

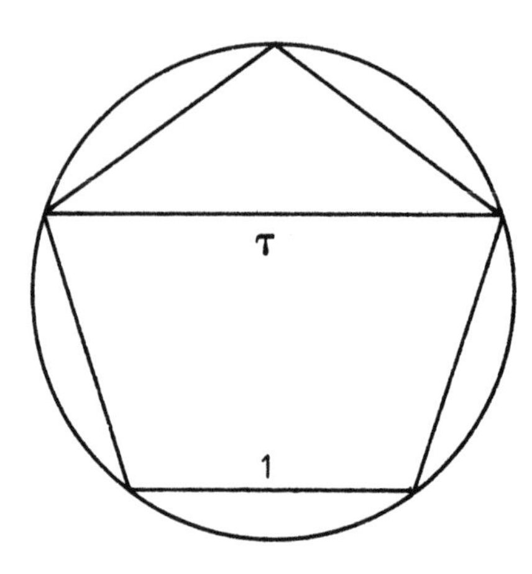

Figure 1.2Q

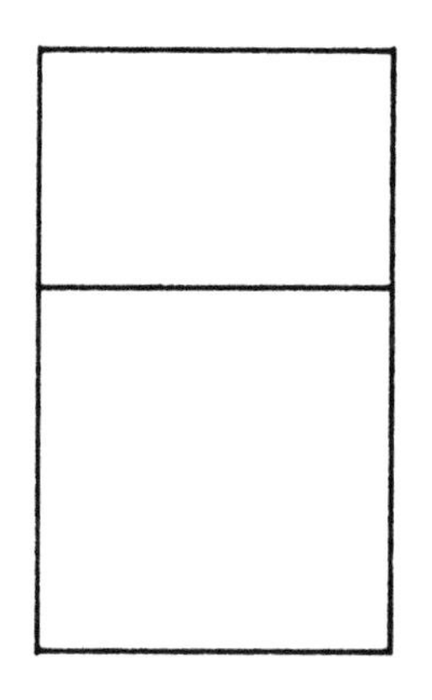

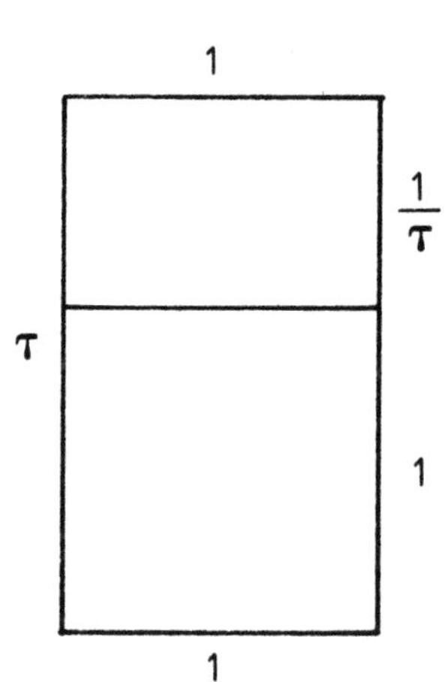

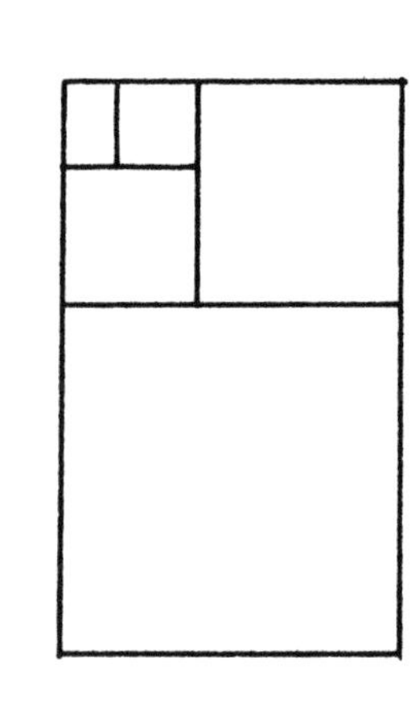

Figure 1.2R

By making quick sketches of such a rectangle you will find that by trial and error you soon acquire a better feeling for this proportion.

From the golden rectangle as shown above, it follows that

$$1 + \frac{1}{\tau} = \tau$$

Multiplying through by τ, this can be re-expressed as the quadratic equation

$$\tau + 1 = \tau^2$$

giving the positive solution

$$\tau = \tfrac{1}{2} + \tfrac{1}{2}\sqrt{5}$$

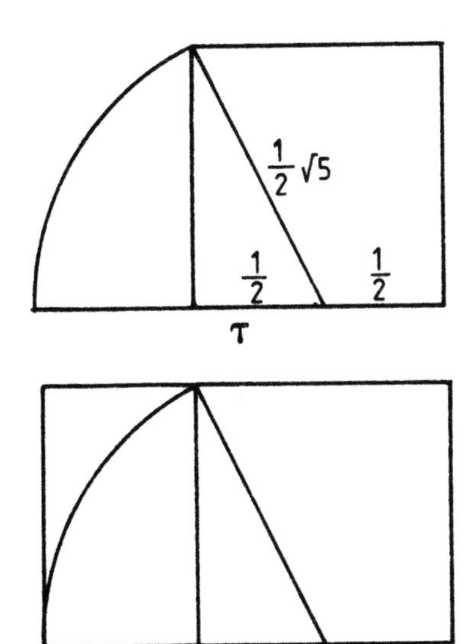

Figure 1.2S

Euclid's method for constructing the golden ratio relies on the theorem of Pythagoras to confirm that the length of the diagonal of a rectangle with sides $\frac{1}{2}$ by 1 is $\frac{1}{2}\sqrt{5}$ (Figure 1.2S).

There are two golden triangles: the 72°–72°–36° triangle and the 36°–36°–108° triangle (Figure 1.2T).

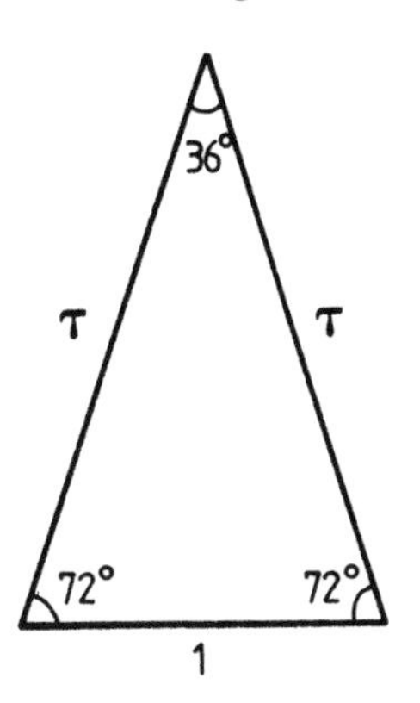

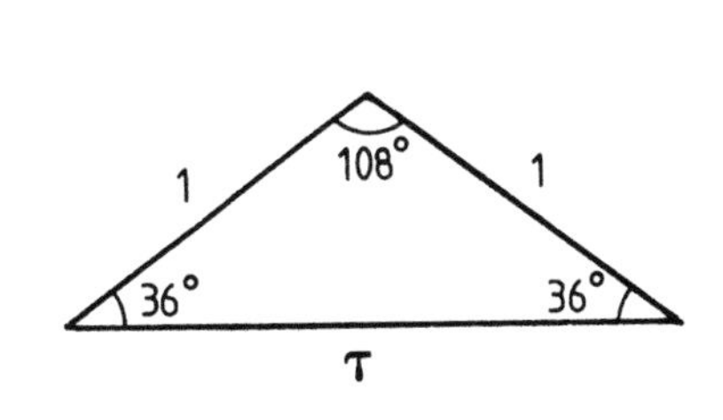

Figure 1.2T

To see that these triangles do have sides in the golden ratio, try to find and compare similar triangles (Figure 1.2U).

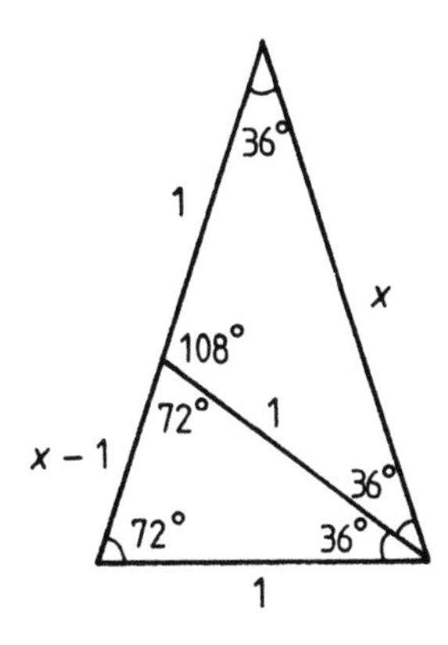

Figure 1.2U

THE REGULAR PENTAGON AND REGULAR DECAGON

Both golden triangles will be found in the regular pentagon, pentagram (five-pointed star) and decagon (Figure 1.2V). It was probably because they were so proud of the particular discovery that the fivefold and tenfold divisions of the circle come by constructing the golden ratio that the early Greek geometers, long before Euclid himself, wore a pentagram as the badge of their profession.

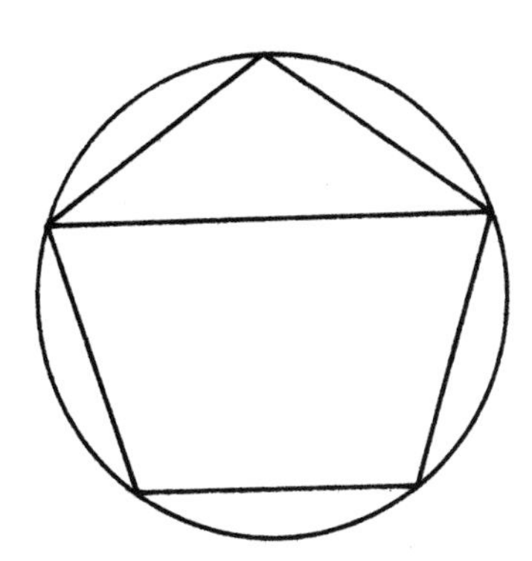

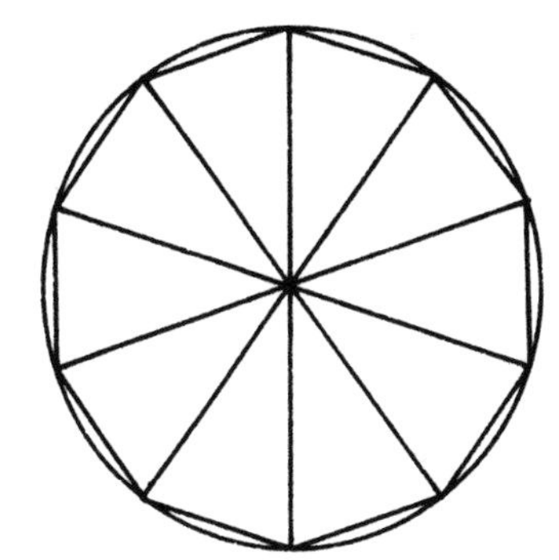

Figure 1.2V

Exercise 2

Construct the following.

1 **a** With respect to a given line, a perpendicular erected at a given point on the line.
b With respect to a given line, a perpendicular through a given point external to the line.
c With respect to a given line, a parallel line through an external point. No short-cuts!

2 Draw any angle. Now, using only compass and rule, construct an equal angle on a given line.

3 **a** Three-fifths of a given line.
b Five-thirds of a given line.
c Three-quarters of a given line.

4 Both types of golden triangle.

5 **a** In a given circle, a regular octagon (eight sides).
b In a given circle, a regular 12-gon.
c In a given circle, a regular 15-gon.
d In a given circle, a regular 30-gon.

6 The centre of a given circle.

7 The tangent at a given point on a circle.

8 About a given point as centre, the circle which touches a given circle. Draw both cases, external and internal respectively, to the given circle.

9 The circle through three given points. The points must not be collinear.

10 Two circles of a given radius that pass through two given points.

11 Without using short-cuts (p. 23), both cases of a common tangent to two given circles, together with their points of tangency.

12 **a** A circle of given radius to touch two given lines, non-parallel.
b A circle of given radius to touch two given circles, when possible.

Reading

Aaboe, 1964; Abbott, 1948; Heath, 1956; Huntley, 1970; Roth, 1948; also any book on Euclidean or elementary geometry.

See Drawings 7–17, pp. 13–22.

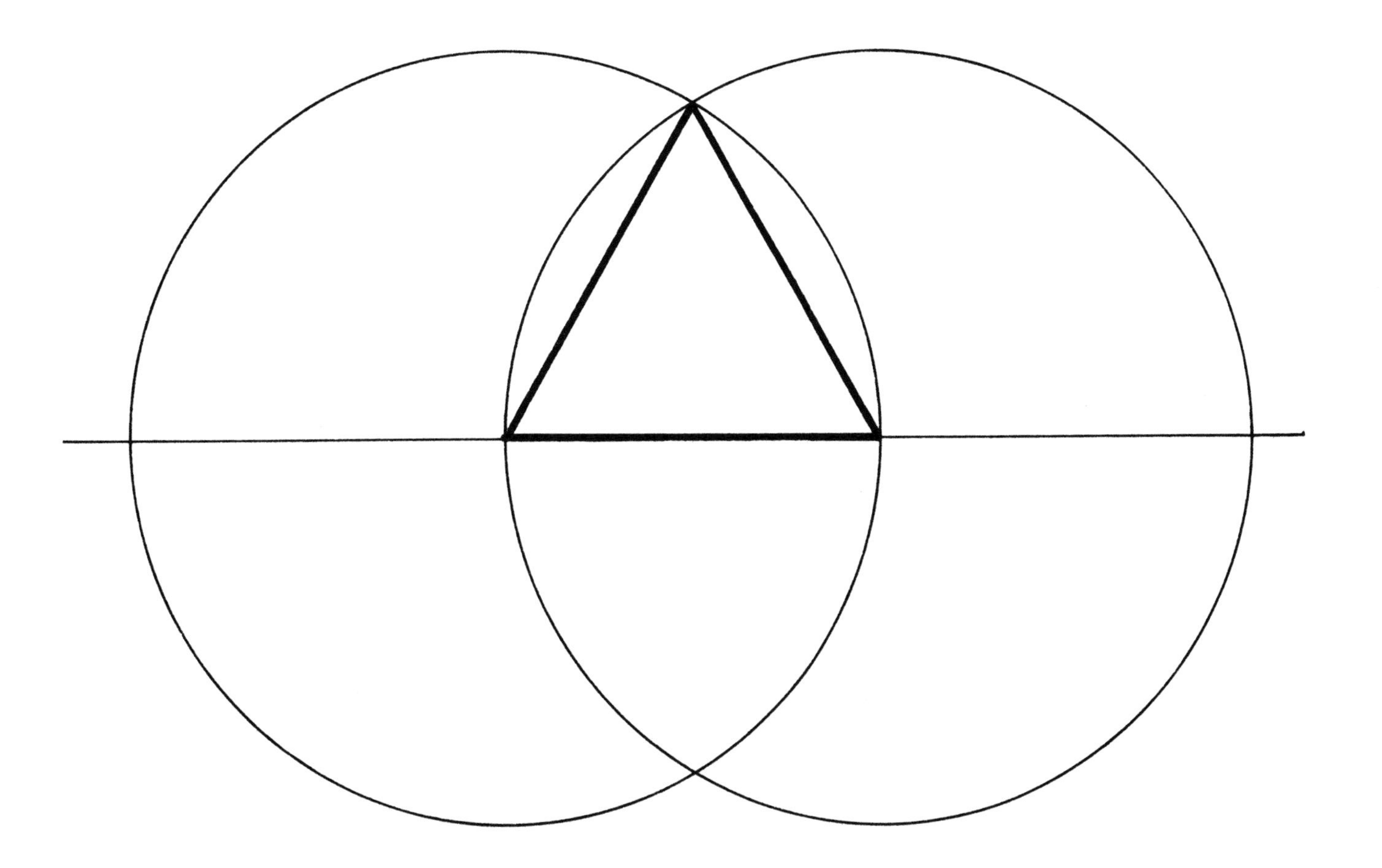

18 Equilateral triangle

1.3 Euclidean approximations

Despite the many constructions which can be done by the Euclidean method, there are even more angles and lengths which cannot be exactly reached in a finite number of steps. Here are some simple examples of such *impossible constructions*.

- **a** To divide any given angle into three equal parts.
- **b** To divide the circle into seven equal parts.
- **c** To divide the circle into nine equal parts.
- **d** To construct the length equal to $\pi = 3.14159\ldots$.
- **e** To construct the cube root of any given length.

The explanations for why these cannot be constructed with compass and rule alone were all discovered in modern times and are too advanced for discussion here. The Greeks themselves must have suspected as much, even though they could not prove it, and turned instead to find other instruments to construct these lengths and angles, as we shall see at the end of the section. The purpose of this section is to present some fluke *approximations* to impossible constructions **b**, **c** and **d** listed above.

The aim is to find a Euclidean construction which, although not exact, is very close to the angle or length in question. You will need to have learnt how to do most of the constructions in the previous section before attempting the drawings in this section. It will also help to have a pocket calculator to hand so as to check the near coincidences. Unlike the drawings of the last chapter which are so universally important and elegant, the present drawings are strange diversions, but they make interesting practice. There is virtually no theory behind them, merely accidental closeness. For that reason, we have no systematic way of finding them. We rely on trial and error to come up with some idea, which we can then check with a calculator. We should not feel surprised to find near misses, since there are so many ways of putting Euclidean constructions together that it would be surprising if there were no simple constructable near misses to the impossible angles and lengths.

The general idea of a Euclidean approximation is quite simple. We look for a lucky near coincidence. Consider, for example, the regular heptagon (7-gon). To divide a circle into exactly seven equal parts would require us to be able to construct the angle of 360°/7 = 51.428571°, which we cannot do. Yet it so happens, and for reasons which are not to do with the heptagon or the number 7, that we can quite easily construct the angle of 51.470 701°:

$$51.470701 - 51.428571 = 0.04213$$

$$\frac{51.470701}{51.428571} = 1.0008$$

$$51.470701 \times 7 = 360.295$$

So the constructable angle of 51.470 701° clearly provides a very good approximating angle for a true one-seventh of a circle. When it is stepped out seven times on its circumference, it completes the 360° with an excess of 0.3° — less than one-thousandth part of a circle. This is small, but it should be visible on a good drawing. With practice, you can achieve a margin of error which is below the level of this discrepancy and so choose to show it accurately. Alternatively, you may wish to try making the minute adjustment (one-seventh of the cumulative discrepancy — any adjustment more than two-sevenths makes the situation worse) required for getting an almost perfect result.

How are these flukes discovered? To give you an idea about how to look for fresh flukes, this chapter presents eight examples. Learn to draw each of them in turn by studying the drawings carefully. The key idea and opening steps of each drawing are described, but you will need to puzzle out the whole drawing for yourself. You will soon learn that each step is one or other of the Euclidean steps introduced in Section 1.2, which should be thoroughly understood before proceeding any further here.

The instructions also contain numerical explanations and assessments which use the theorem of Pythagoras and the sine, cosine and tangent ratios. These are the subjects of Sections 4.1 and 4.4. Nevertheless, you should be able to follow the instructions for *doing* the drawings without this knowledge (arithmetic and trigonometry), and readers are encouraged to learn these drawings as compass exercises.

The eight examples are as follows.

- Drawings **19–21**, three ways of approximating the regular heptagon.
- Drawings **22–24**, two ways of approximating the regular enneagon (nine sides). (Drawings **23** and **24** are different versions of the same principle of construction.)
- Drawings **25–27**, two ways of squaring or rectifying the circle.
- Circling the square (the full drawing is left for you to complete).

They are arranged in order of difficulty, beginning with the easy and leading by degrees to the not so easy.

THE REGULAR HEPTAGON

Drawings **19–21** show three different approximations to the regular heptagon (Figure 1.3A), each with its own measure of discrepancy. The basic idea in each of the three methods is to construct an angle which is very close to one-seventh of a circle:

$$360°/7 = 51.428571°$$

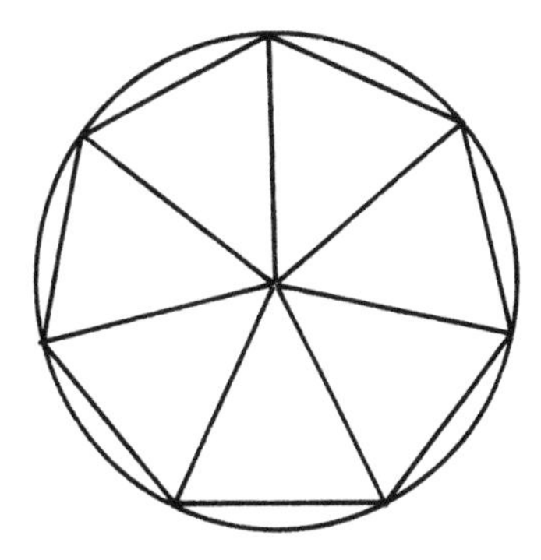

Figure 1.3A

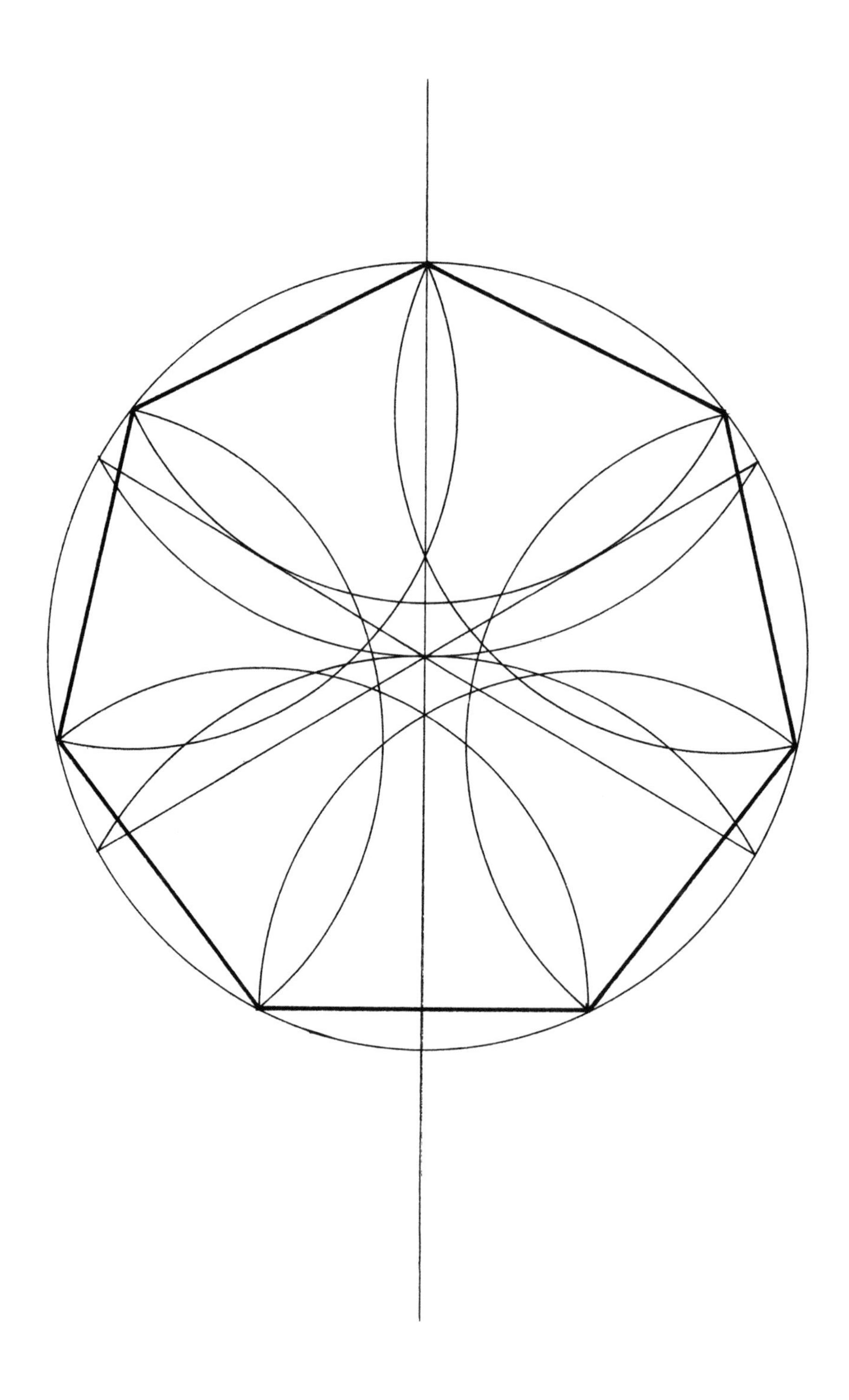

19 Heptagon: method I

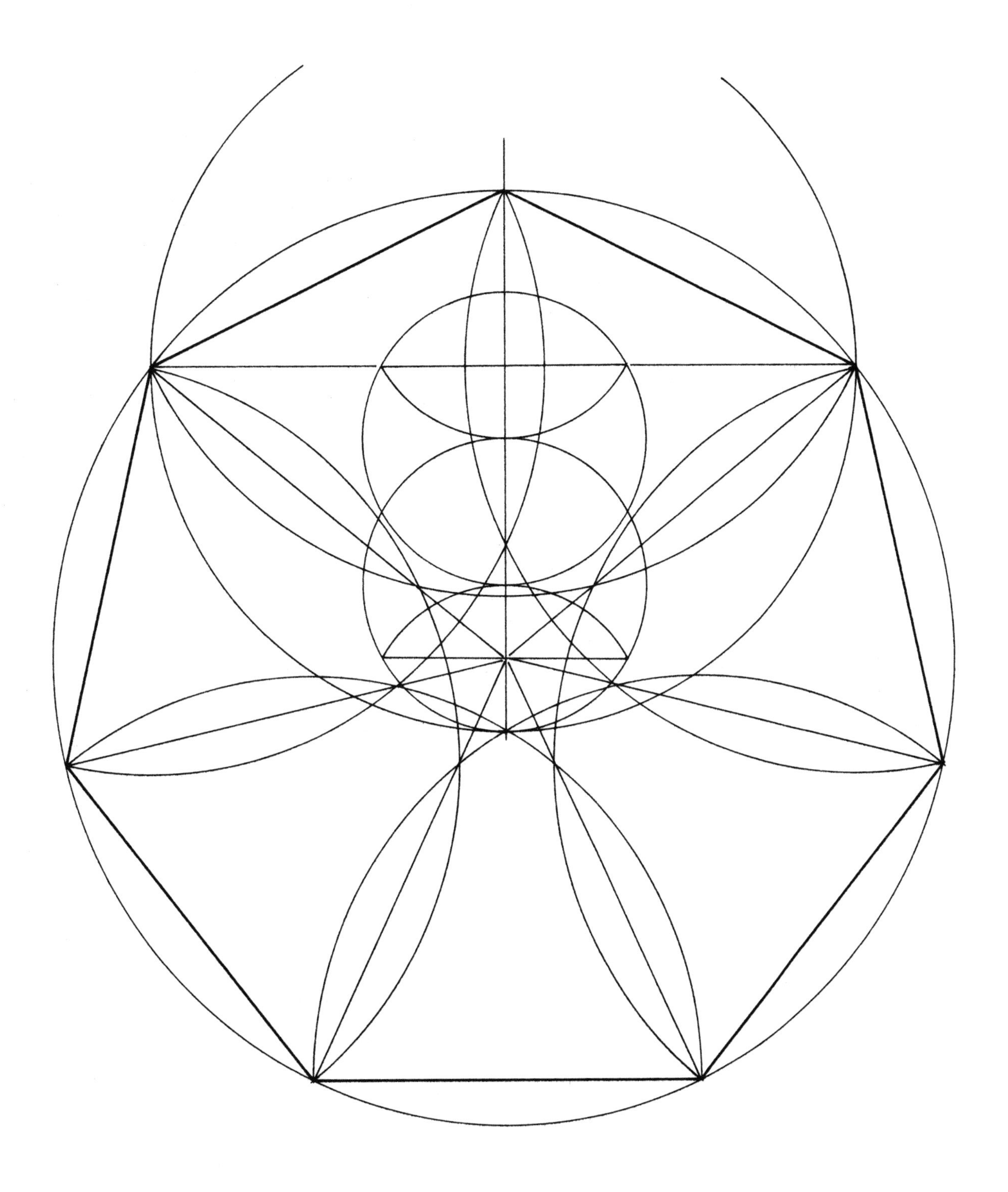

20 Heptagon: method II

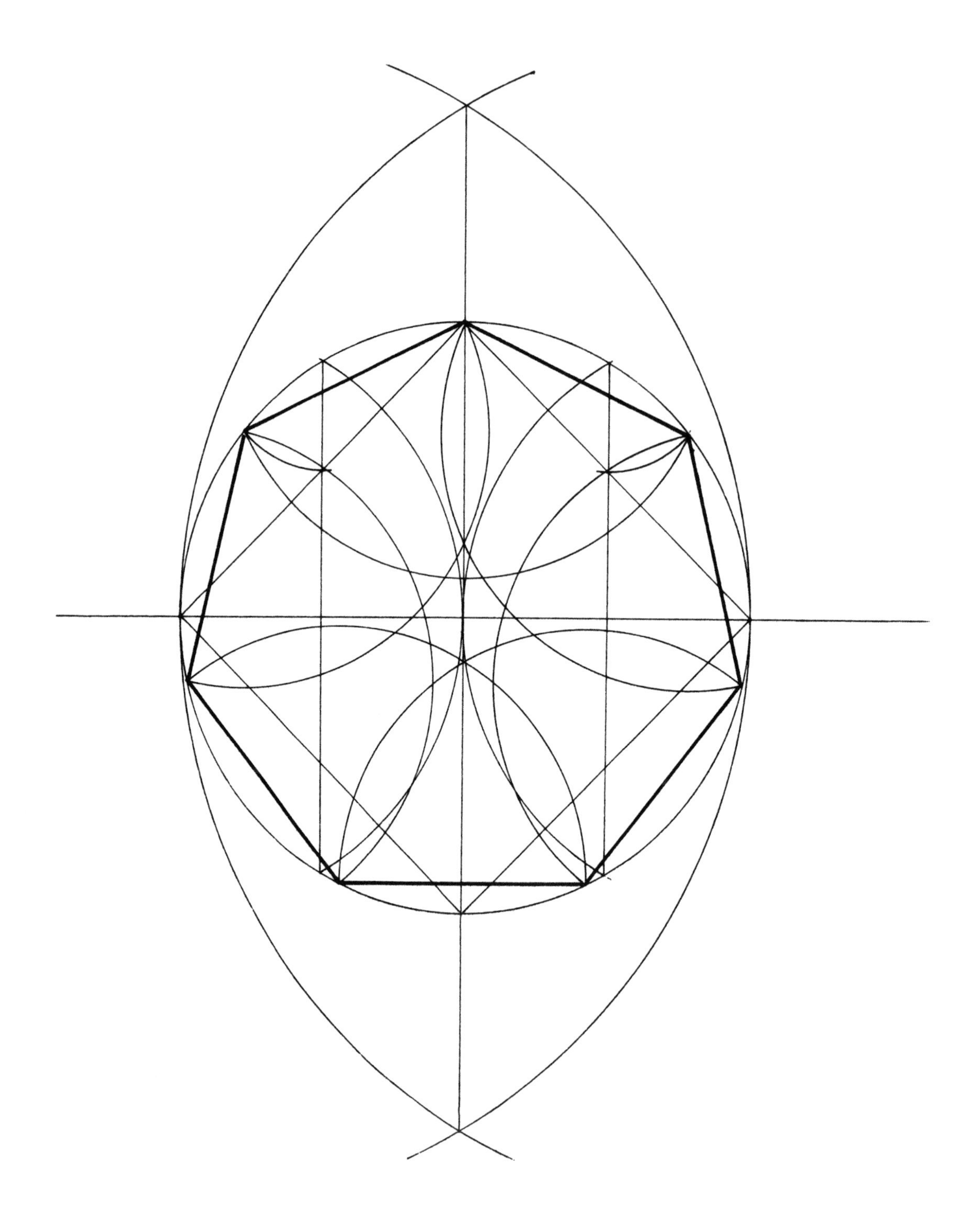

21 Heptagon: method III

Method I: to construct 51.317 813°

In the circle which is to be divided into seven equal parts, construct an equilateral triangle, and take half of the side of the triangle to be the side of the heptagon (Figure 1.3B).

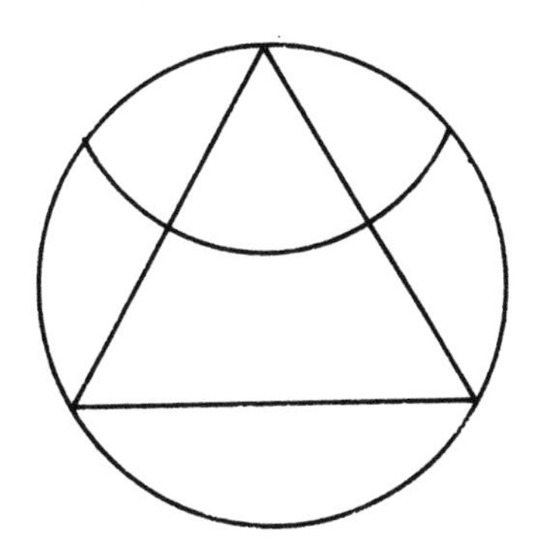

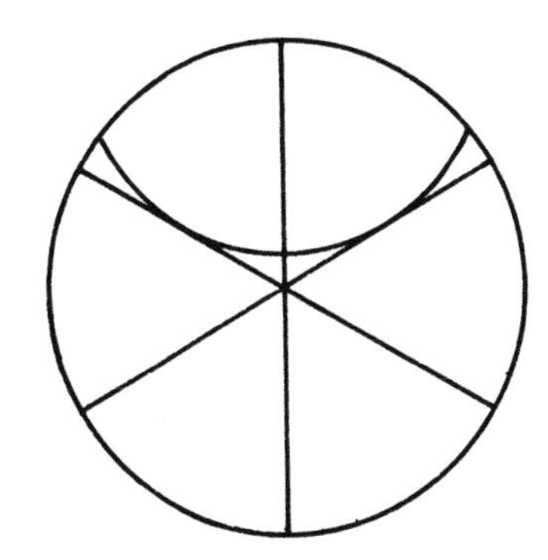

Figure 1.3B

Numerical assessment of method I

The chord of 120° of a unit circle equals $\sqrt{3}$. The angle whose chord is $\frac{1}{2}\sqrt{3}$ = 51.317 813°.

$$51.428571 - 51.317813 = 0.110758$$

$$\frac{51.317813}{51.428571} = 0.998$$

$$51.317813 \times 7 = 359.22$$

The discrepancy of this approximation is about two parts per thousand, or about four-fifths of a degree short in seven steps around the circle.

Method II: to construct 51.340 192°

Construct a right-angled triangle whose perpendicular sides are 5 and 4 respectively (Figure 1.3C). Take the angle between the hypotenuse and the shortest side to be one-seventh of a circle.

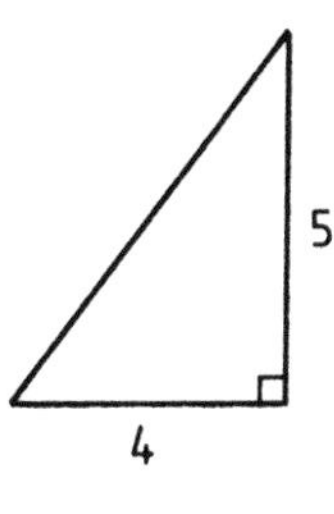

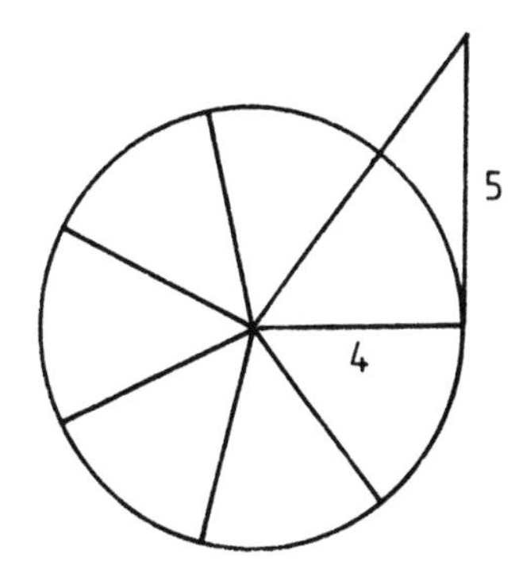

Figure 1.3C

Numerical assessment of method II

The angle whose tangent is 5/4 = 51.340 192°

$$51.428571 - 51.340192 = 0.088$$

$$\frac{51.340192}{51.428571} = 0.998$$

$$51.340192 \times 7 = 359.38$$

This approximation is very much the same as that in method I, slightly more than three-fifths of a degree (instead of nearly four-fifths) short of 360° in seven strides of the compass.

Method III: to construct 51.470 701°

In the unit circle to be divided into seven equal parts, construct a six-petalled flower, as shown in Figure 1.3D. Draw an arc centred on one petal point to touch the neighbouring petals and to cut the circle. Take the angle between this point and a further petal point, as shown, to be two-sevenths of a circle. Bisect it to obtain the side of the heptagon.

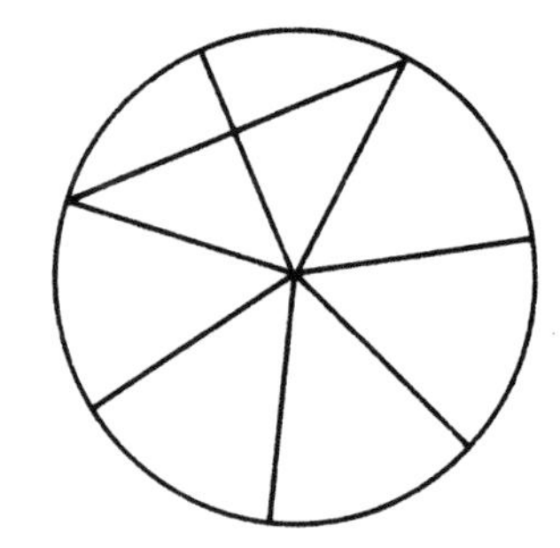

Figure 1.3D

Numerical assessment of method III

The arc centred on one petal point to touch a neighbouring petal equals $\sqrt{3} - 1$. The angle whose chord is $\sqrt{3} - 1 = 42.941403°$.

$$60° + 42.941403° = 102.941403°$$

$$102.941403 \times \tfrac{1}{2} = 51.470701$$

This is the closest approximation of the three, exceeding 360° in seven strides of the compass by 0.3°. The figures for this approximation were given earlier (p. 35).

THE REGULAR ENNEAGON

Drawings **22–24** show two different methods for approximating an exact division of the circle by 9 (Figure 1.3E).

$$360°/9 = 40°$$

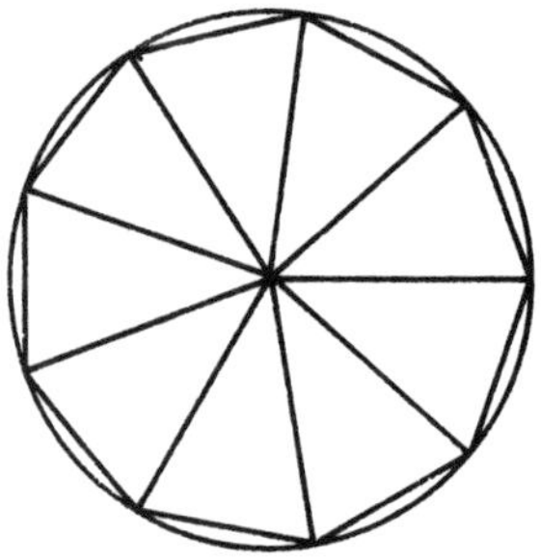

Figure 1.3E

Method I: to construct 39.805571°

Construct a right-angled triangle whose perpendicular sides are 5 and 6 (Figure 1.3F). Take the angle between the 6 and the hypotenuse to be one-ninth of a circle.

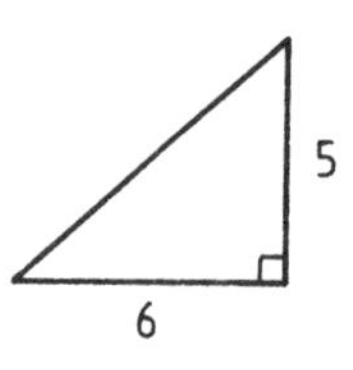

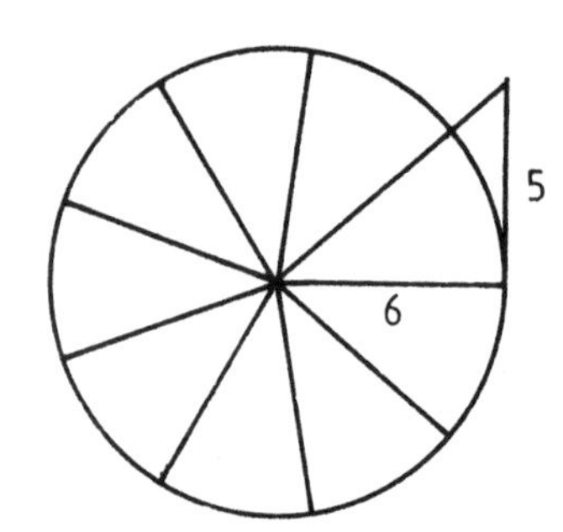

Figure 1.3F

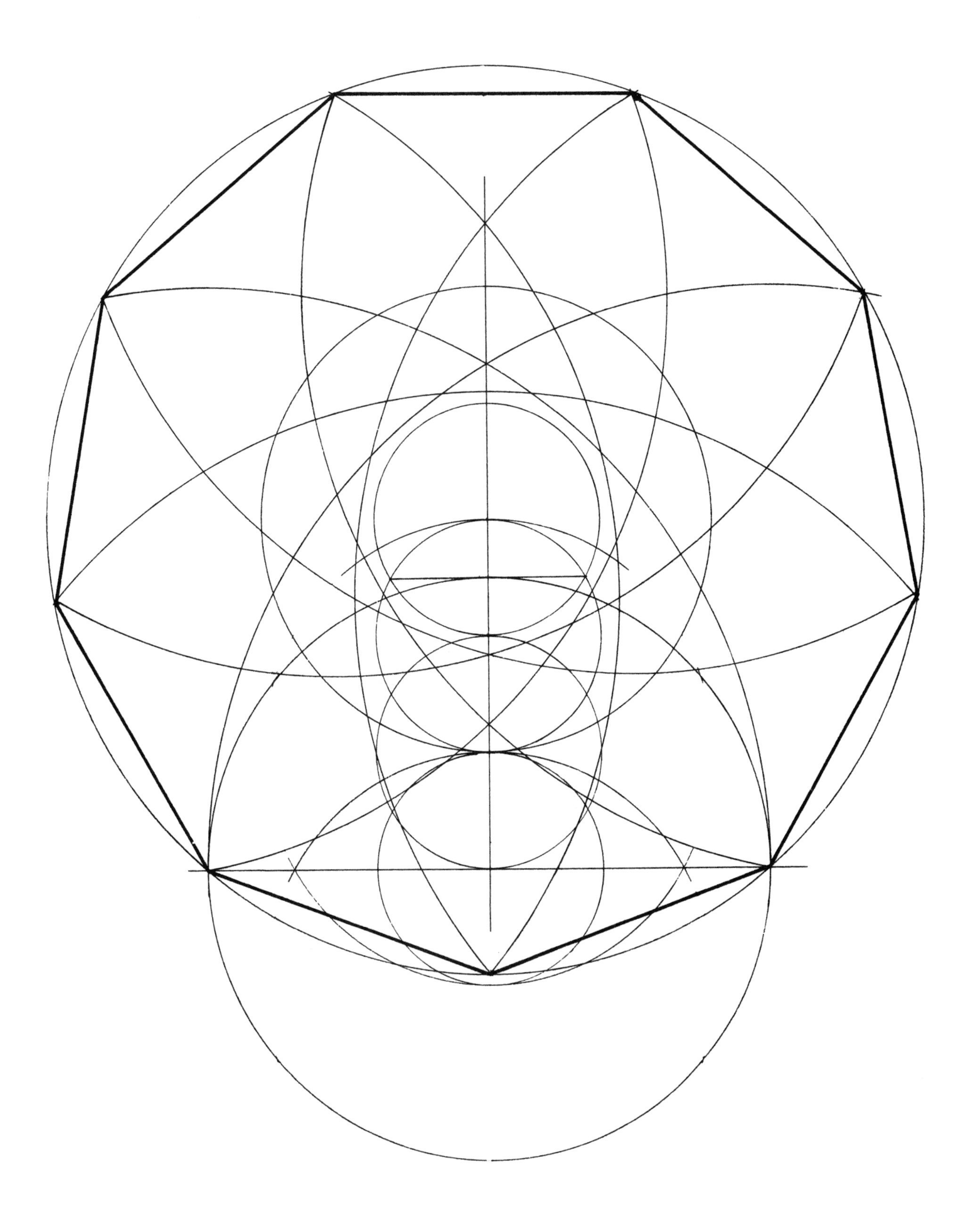

22 Enneagon: method I

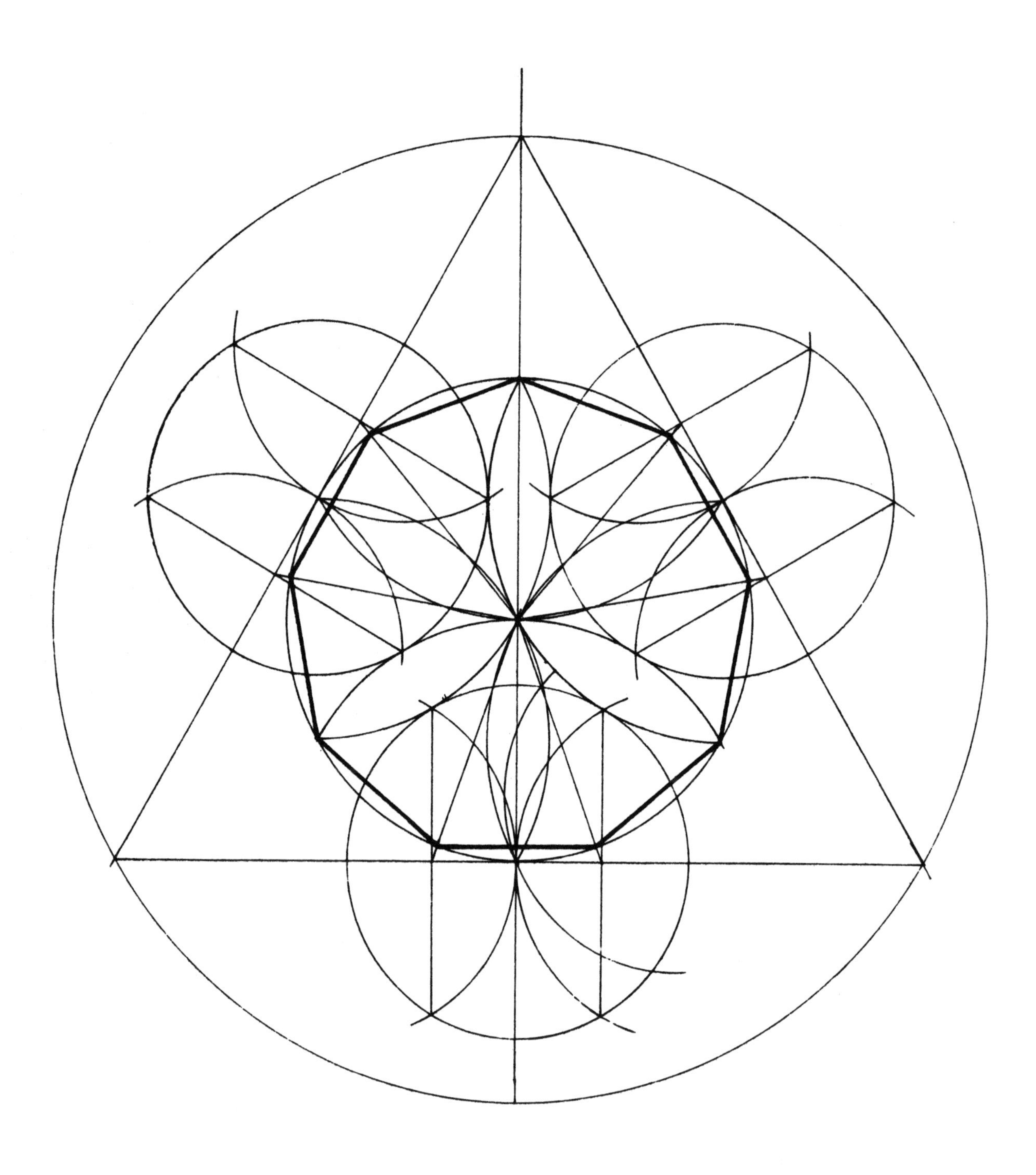

23 Enneagon: method II, version 1

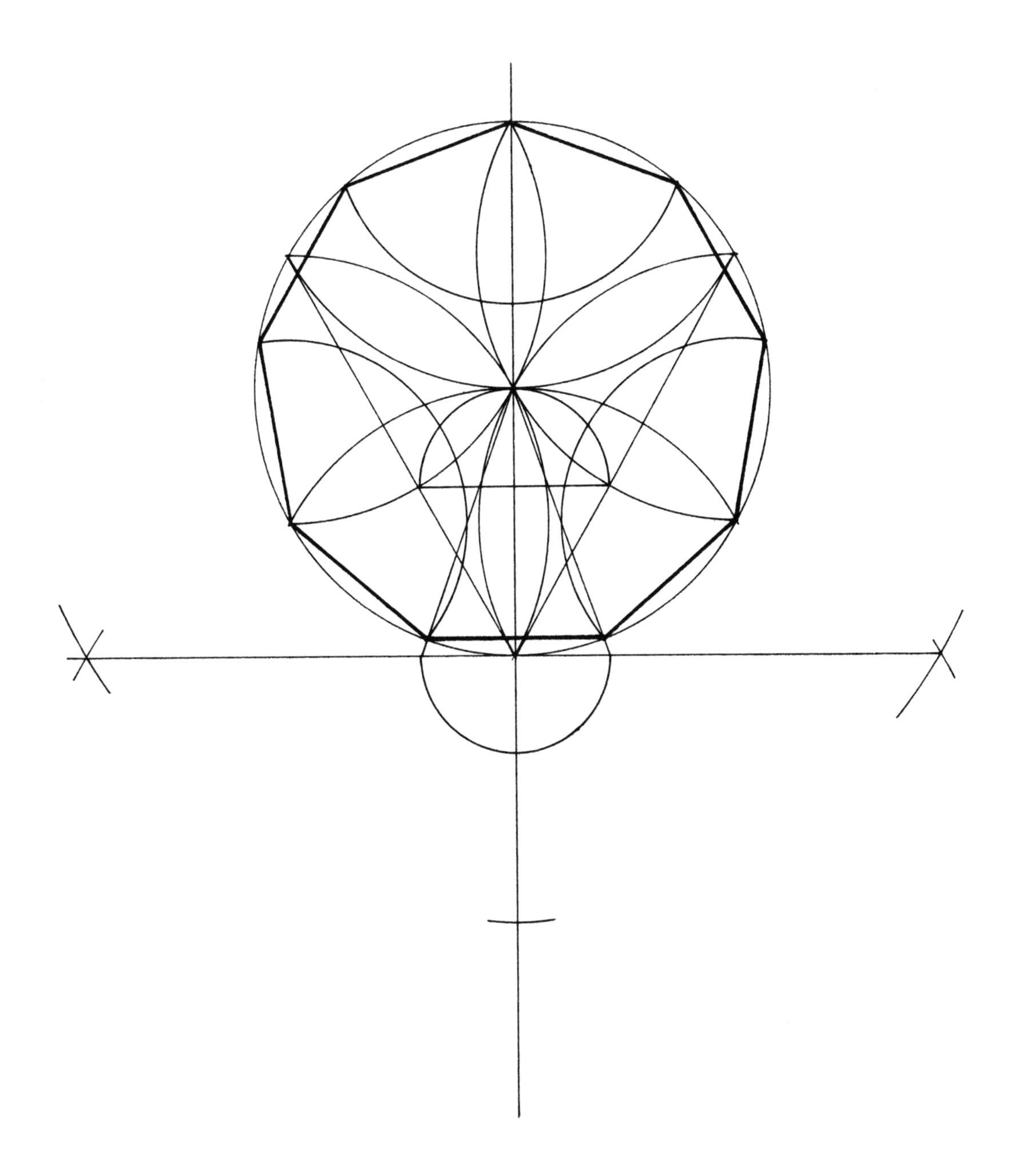

24 Enneagon: method II, version 2

Numerical assessment of method I

$$\arctan(5/6) = 39.805571°$$

Nine strides with this angle falls short of 360° by $1\frac{3}{4}°$, which should be clearly visible.

Method II: to construct 40.207 819°

Construct a right-angled triangle with sides 2 and $\sqrt{3} - 1$ and take the sharp angle to be 20°. In the circle to be divided into nine equal parts, the diameter provides the measure of 2 and the distance between one petal point and a neighbouring petal as the measure of $\sqrt{3} - 1$, as in the approximate heptagon method III (Figure 1.3G).

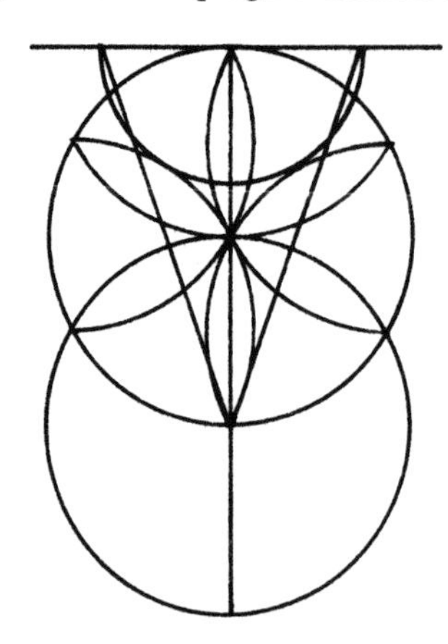

Figure 1.3G

Note that a threefold division of the circle to be divided into nine parts avoids having to step nine times around the circle with the chord of 40°, giving a more reliable construction.

Numerical assessment of method II

The angle whose tangent is

$$(\sqrt{3} - 1)/2 = 20.103909°$$

$$20.103909° \times 2 = 40.207819°$$

$$\frac{40.207819}{40} = 1.0051955$$

This means that, of the nine angles in the constructions, six are $\frac{1}{2}$% too large and three are 1% too small.

SQUARING THE CIRCLE

How far is it round a circle? More than three diameters, approximately $3\frac{1}{7}$ diameters, but how far exactly? We call this ratio π, and in Section 3.2 we shall see how

$$\pi = 3.141\ 592\ 7\ \ldots$$

was eventually solved, in terms of numerical calculation. The problem put in geometric terms is this: construct a length equal to π (Figure 1.3H).

It turns out that this cannot be done with compass and rule alone. However, because no one until modern times was able to prove that there can be no Euclidean construction of π, many people continued to try. The problem became famous in two versions (Figure 1.3I): rectifying the circle, i.e. constructing a square equal in perimeter to a given circle (Drawings **25** and **26**); squaring the circle, i.e. constructing a square equal in area to a given circle (Drawing **27**).

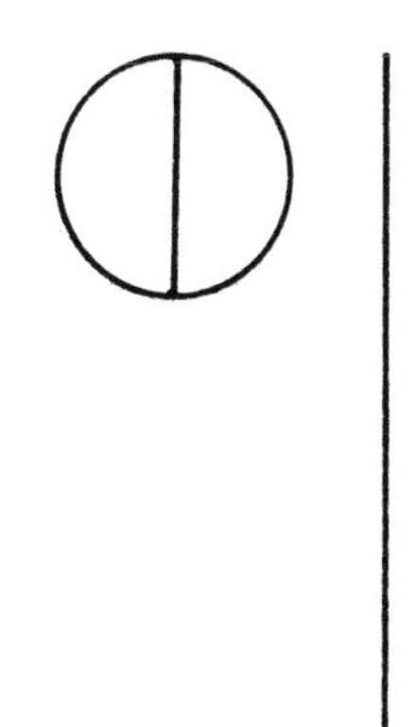

Figure 1.3H

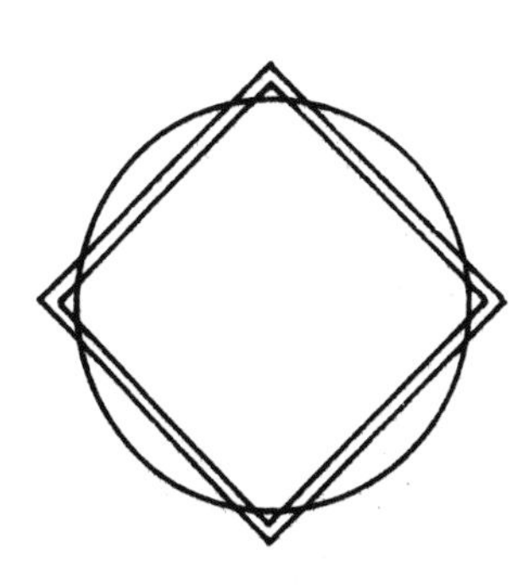

Figure 1.3I

It is easy to work out that, if the circle in question has a radius equal to 1, then the side of the rectifying square must be $\pi/2$, and that of the squaring square must be $\sqrt{\pi}$. This means that the two versions of the problem are equivalent. For, if we could construct the one, we could therefore construct the other. This is because we know how to construct the square or the square root of any given length by Euclidean construction.

Drawing **25** shows method I for rectifying the circle, and drawing **27** shows how, by taking the further step of constructing a square root, this gives an approximate squaring of the circle. The radii of the inscribed circles to the rectifying square and the squaring square are $\pi/4$ and $\frac{1}{2}\sqrt{\pi}$ respectively. Drawing **26** shows method II for rectifying the circle.

Method I: to construct 3.1416408

This method is quite complicated, as you can see in Figure 1.3J. The idea is to construct three-tenths of $1 + \tau$, where τ (= 1.618034) is the golden ratio. Take this to be your approximation for $\pi/4$, the radius of the circle circumscribed by the rectifying square.

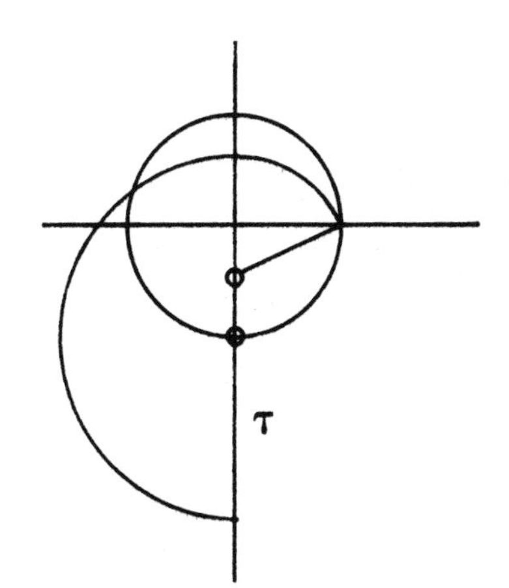

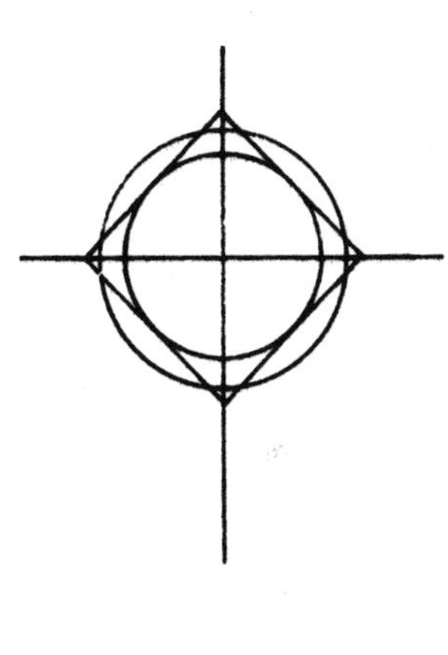

Figure 1.3J

To square the circle, construct the square root of the above-mentioned radius to give the radius of the circle circumscribed by the squaring square (Drawing **27**).

Numerical assessment of method I

$$4 \times \left\{\frac{3}{10}(1 + \tau)\right\} = 3.1416408$$

$$\frac{3.1416408}{3.1415927} = 1.0000153$$

The discrepancy here is very small indeed, exceeding π by only 15 parts per million.

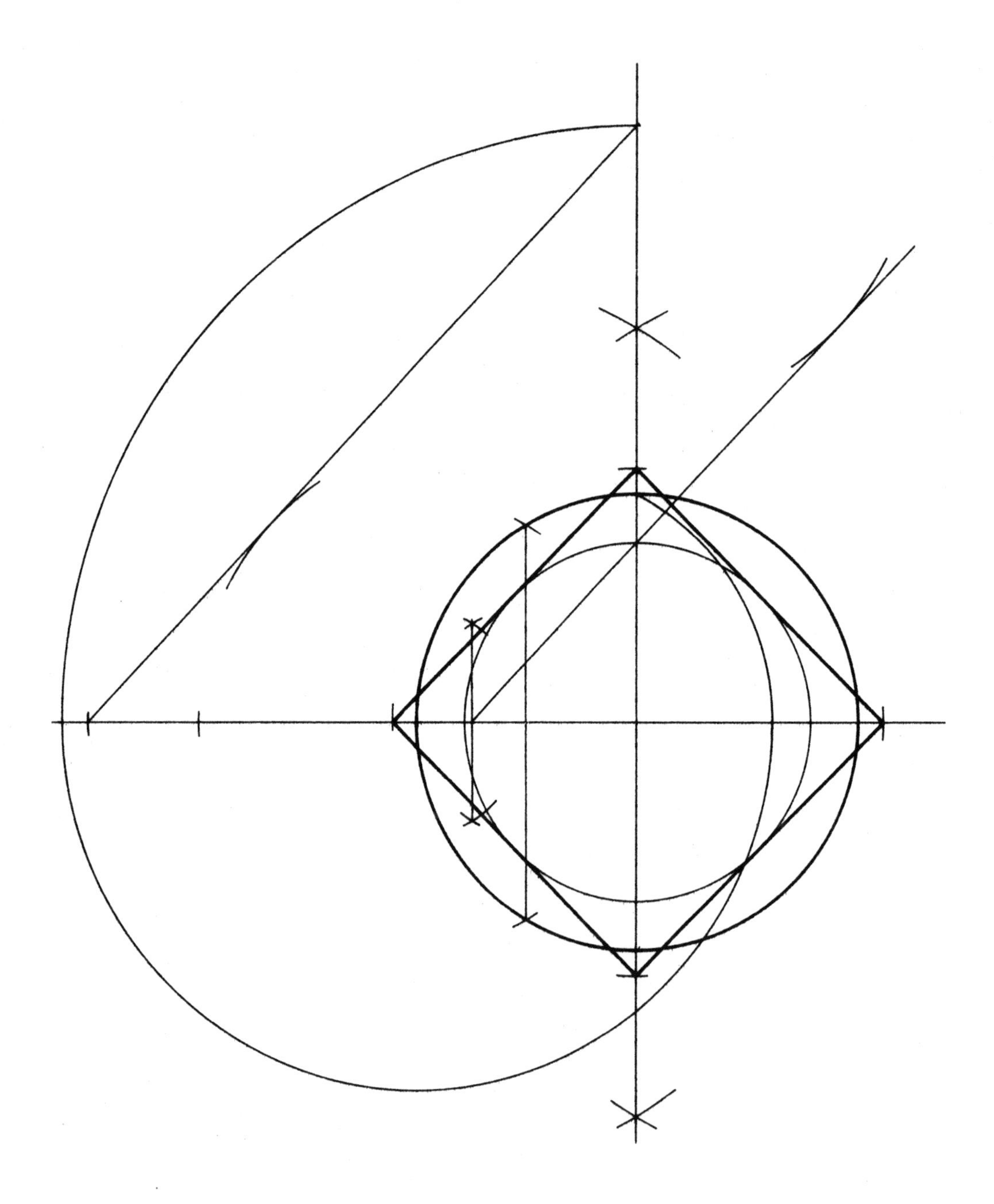

25 Rectifying the circle: method I

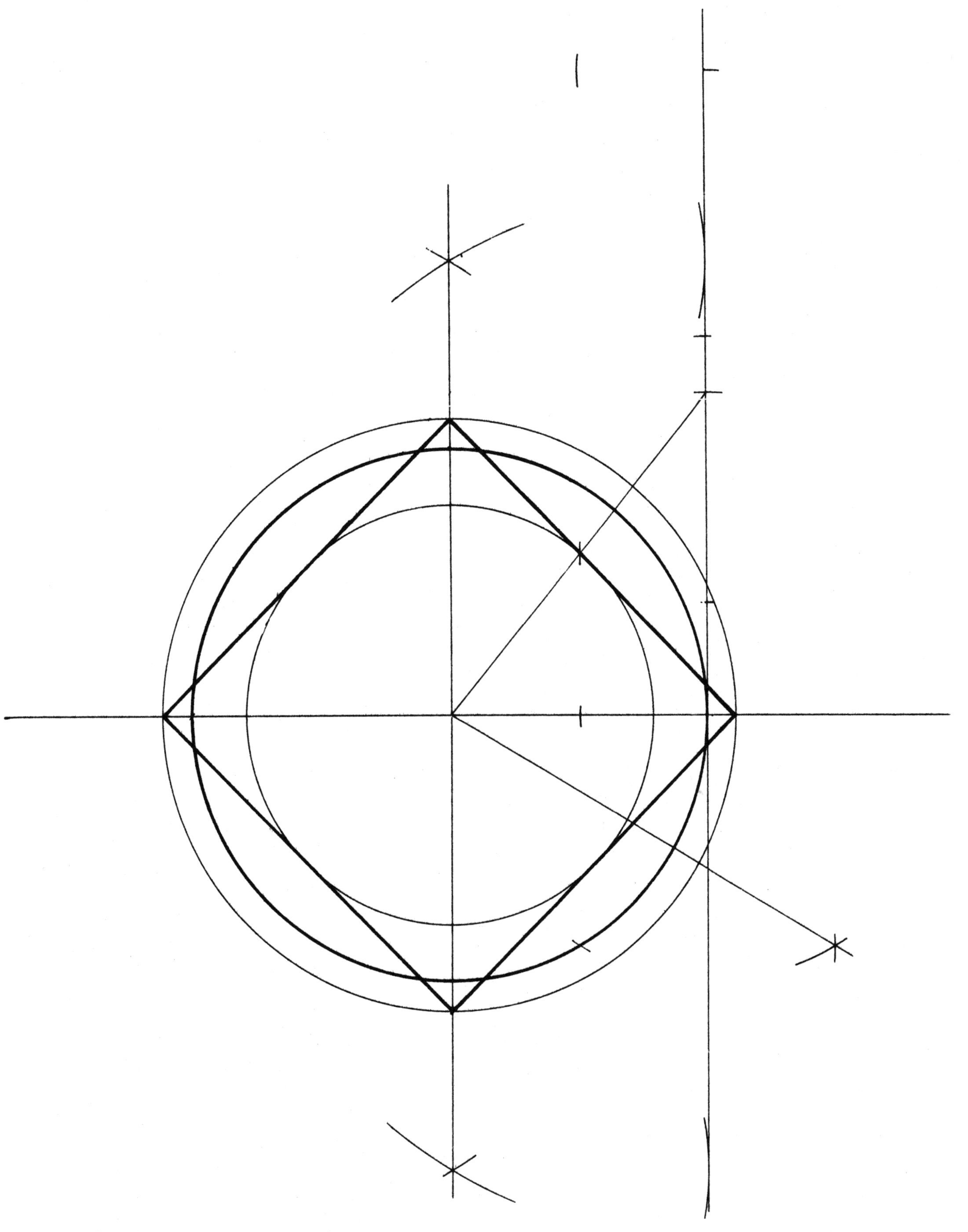

26 Rectifying the circle: method II

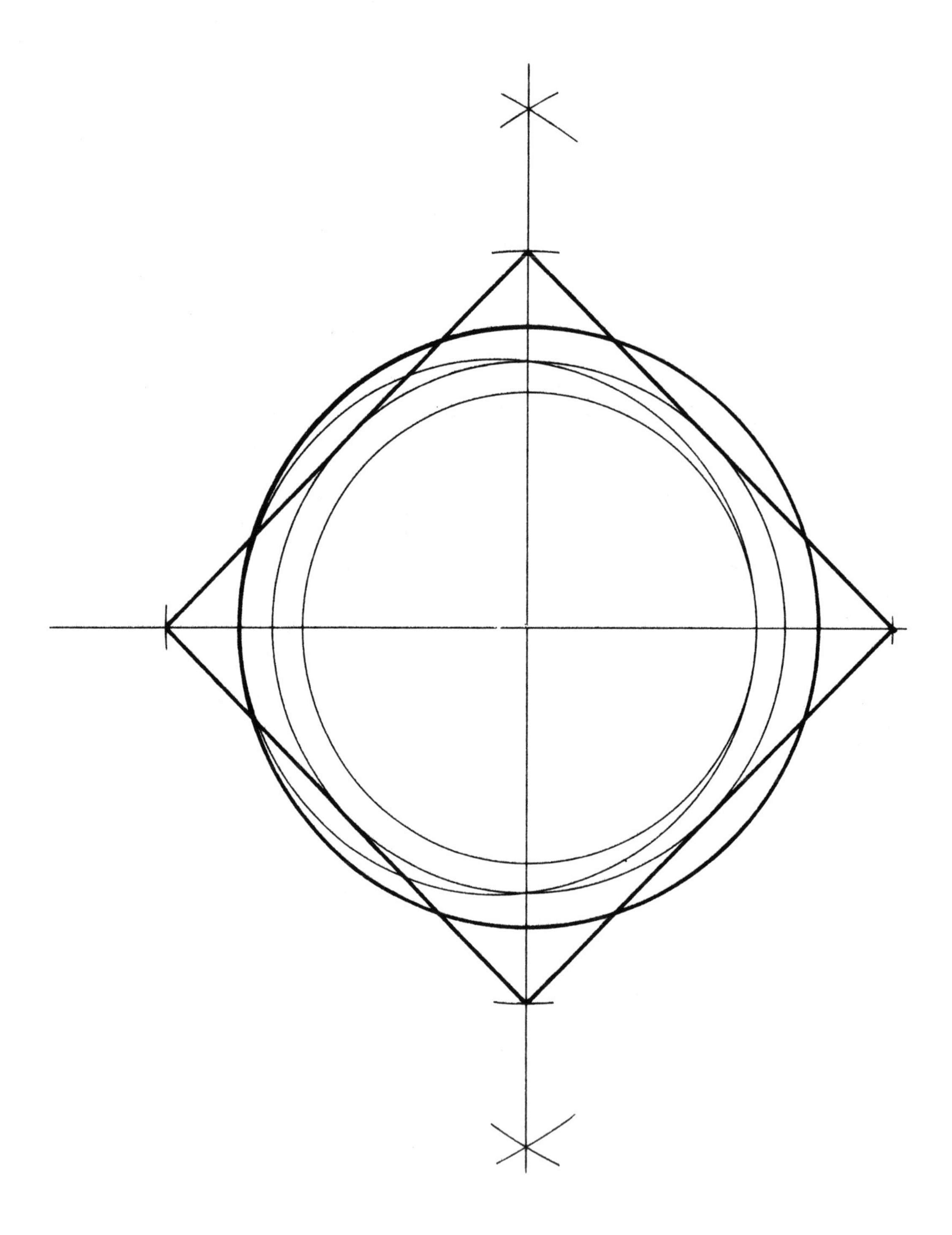

27 Squaring the circle: extension of method I

Method II: to construct 3.141 533 3

Again, this method is quite difficult to follow. Starting with the circle to be squared, assumed to have a radius equal to 1, construct the tangent of 30°, as shown in Figure 1.3K. Next extend this line so that a line is constructed which has a length equal to 3−tan 30°, as shown. Take the hypotenuse of a right-angled triangle whose perpendicular sides are 2 and 3−tan 30° to be the length π.

To construct $\pi/4$, first bisect the length 3−tan 30° and then connect this mid-point to the centre of the circle as shown. Take half of this hypotenuse to be $\pi/4$.

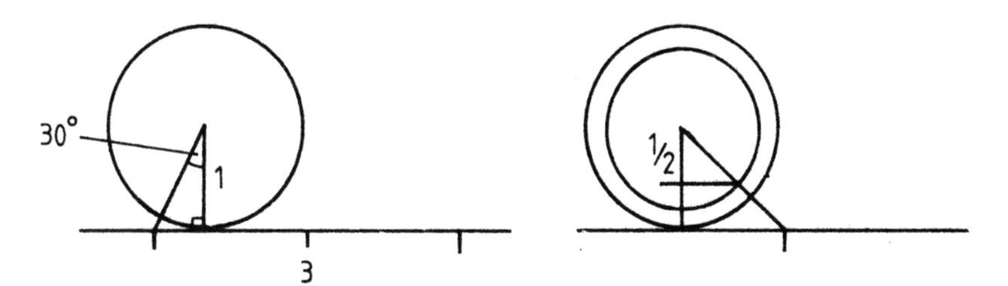

Figure 1.3K

Numerical assessment of method II

The hypotenuse of a right-angled triangle having perpendicular sides 2 and 3−tan 30° is

$$\sqrt{\{4 + (3-\tan 30°)^2\}} = 3.1415333$$

$$\frac{3.1415333}{3.1415927} = 0.99998$$

The discrepancy here is 20 parts per million. Again, this is very small indeed.

CIRCLING THE SQUARE

This is the reverse of our previous problem of squaring a circle. Here we start with the square and try to construct a circle of equal perimeter (circumference) to the square. Suppose that the circle inscribed in the given square has a radius equal to 1, what will the radius r of the circling circle be (Figure 1.3L)? Clearly the perimeter of the given square is 8, so that the circumference of the circling circle must be 8. Therefore,

$$2\pi r = 8$$

giving

$$r = \frac{4}{\pi} = 1.2732395$$

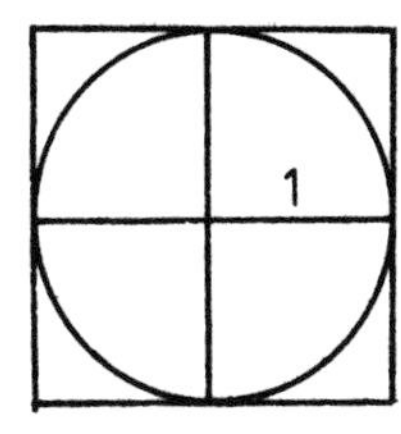

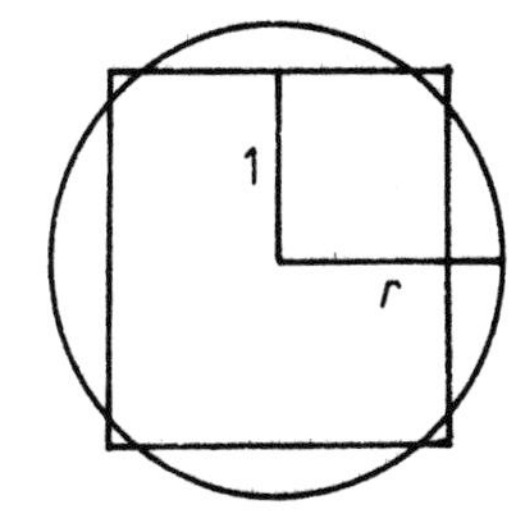

Figure 1.3L

Method: to construct 1.272 019 7

First inscribe the circle in the given square and, by drawing the diagonals of the square, obtain two perpendicular diameters in the inscribed circle. Next draw the two circles on one of the diameters as shown, each of radius half that of the inscribing circle and centred to touch the centre of the inscribing circle (which is also the centre of the square). Then, by centring the compass at the ends of the other diameter, draw arcs to touch the inner pair of smaller circles as shown in Figure 1.3M.

  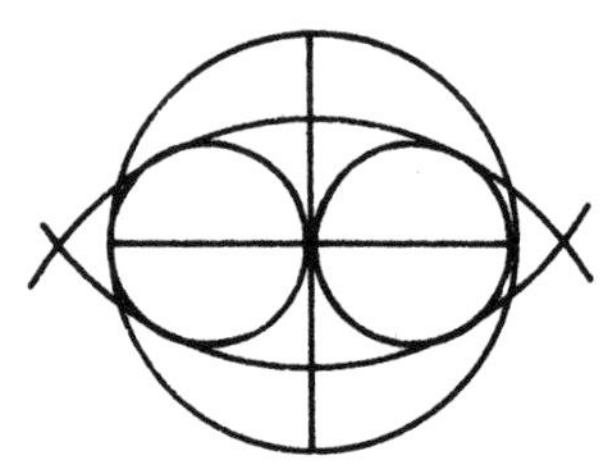

Figure 1.3M

You may recognise that the radius of this arc is the golden ratio. The two points where these arcs meet gives us the diameter of the circle we seek: the circle which very nearly circles the square.

Numerical assessment

Assuming that the inscribing circle of the given square has a radius equal to 1, then the two large arcs have radii equal to 1.618 034, the golden ratio. By Pythagoras' theorem we thus learn that the radius of our approximating circle is $\sqrt{\tau}$:

$$\sqrt{\tau} = 1.272\ 019\ 7$$

$$\frac{4}{\pi} - \sqrt{\tau} = 0.001\,22$$

$$\frac{4}{\pi} \div \sqrt{\tau} = 1.000\,959$$

This is a discrepancy of 1 part in a 1000, which is quite close and certainly very pleasing as a construction.

EPILOGUE ON UNCONSTRUCTABLE ANGLES AND LENGTHS

Although none of the angles and lengths that we have been looking at in this chapter is constructable by compass and rule alone, Greek mathematicians found other devices for constructing them exactly. Archimedes (287–212 BC) in particular demonstrated several methods for trisecting an angle or squaring the circle. Here, by way of example, is one of his solutions for trisecting any given angle (Figure 1.3N).

AOB is the angle to be trisected. Let ED be a rigid rod of length OB whose end point E is free to move on OA while its end point D is free to move on the circle centred on O of radius OB. By moving the rod until EDB is a straight line, obtain the angle DEO = $\frac{1}{3}$AOB.

With ED = OB and EDB collinear, AEB = AOB.

Reading

Aaboe, 1964; Critchlow, 1976; Lawlor, 1982; Steinhaus, 1983.

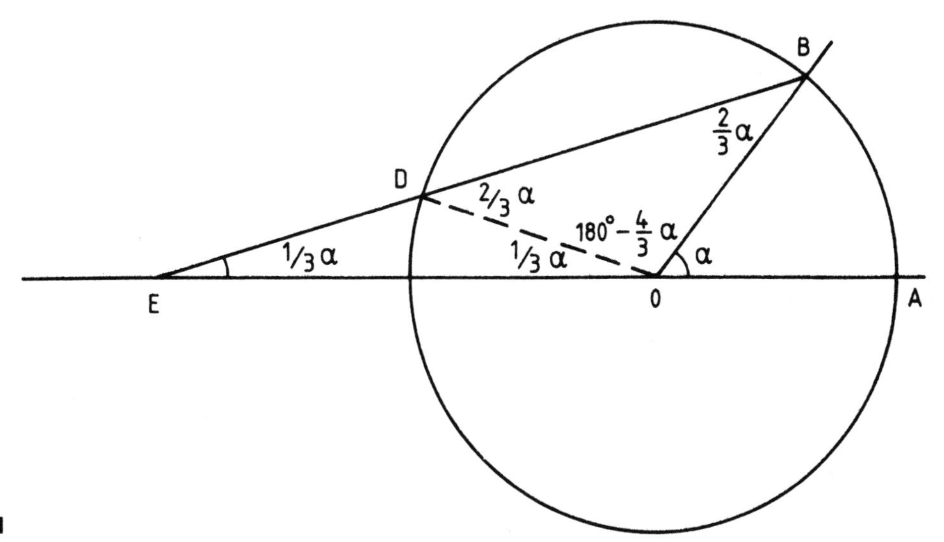
B
$\frac{2}{3}\alpha$
D
$2/3\ \alpha$
$180° - \frac{4}{3}\alpha$
$1/3\ \alpha$
$1/3\ \alpha$
α
E
0
A

Figure 1.3N

1.4 Gauss extends Euclid

CONSTRUCTING A 17-GON

Euclid tells us how to divide the circle exactly into the following numbers of equal parts: primarily,

2, 3 and 5

giving also

15 (by subtracting $\frac{1}{3}$ from $\frac{2}{5}$ to give $\frac{1}{15}$)

and then

4, 8, 16, ...
6, 12, 24, ...
10, 20, 40, ...
30, 60, 120, ...

by continued angle bisection.

Is this a complete list of exact whole number divisions of a circle possible by Euclidean construction? For 2000 years after Euclid's time, no mathematician seems to have suspected that there might be other possibilities, until Karl Friedrich Gauss (1777–1855) discovered that an exact division by 17 is also possible.

The construction of the regular 17-gon (Drawing **28**) shown here is due to Richmond. It is quite tricky to complete accurately, because it involves more than 30 steps to obtain the first one-seventeenth of a circle. However, with practice, it is just about manageable and includes a nice variety of Euclidean constructions (Section 1.2): perpendiculars, line bisections, angle bisections and parallels.

Method (Figure 1.4A)

- **a** Construct the circle which is to be divided into 17 equal parts, and in it construct two perpendicular diameters. Find by repeated line bisection the point A on one of the diameters which is one-quarter of the radius from O, the centre.
- **b** Using the point A as the centre, draw the arc XY as shown.
- **c** Find by repeated angle bisection the quarter of this arc, as shown, and hence the point B of intersection with the other diameter.
- **d** Draw a line at 45° to AB to cut the same diameter at C. (Can you see how to construct this 45° line in three steps? Clue: angle OXZ = 45°.)
- **e** Construct the circle, taking CX as diameter to obtain the points of intersection D and E.
- **f** Construct the circle of radius BD and centre B.
- **g** Construct the two tangents to this small circle parallel to DE to cut the original circle.
- **h** Take the five points of the circumference as shown to be vertices 1, 4, 6, 13 and 15. The other 12 vertices of the regular 17-gon are now easily found.

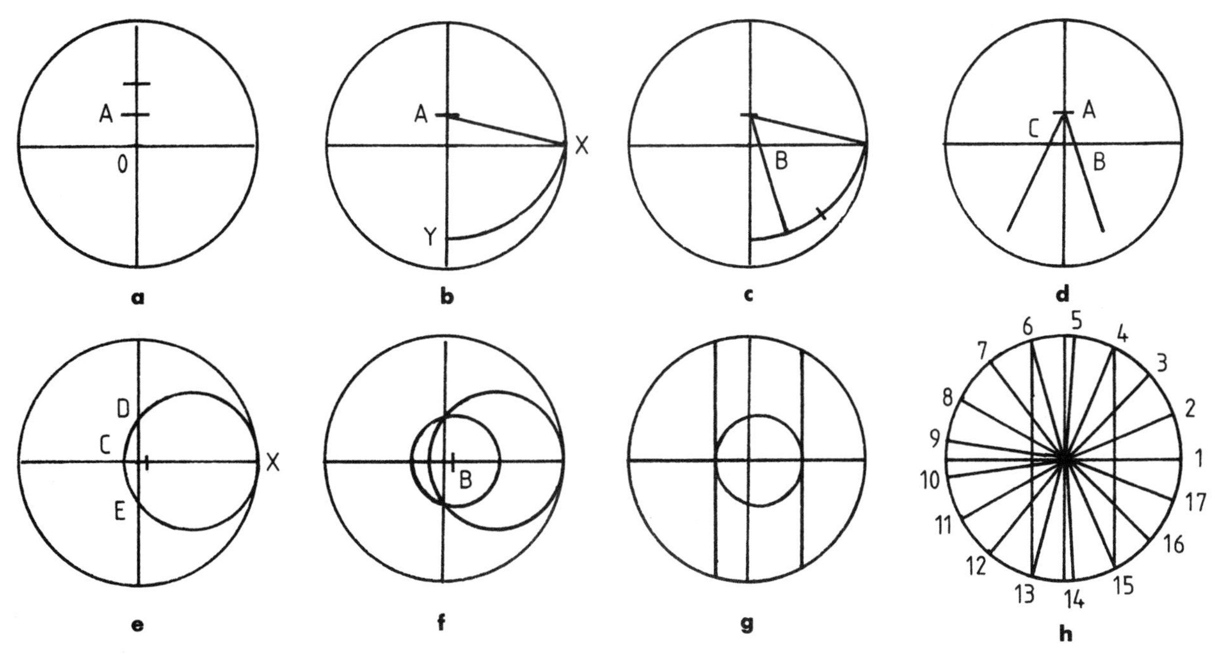

Figure 1.4A

FERMAT PRIMES

These are all the prime numbers which can be written in the following form:

$$2^{2^n} + 1$$

They are named after Pierre de Fermat (1601–1665) who guessed wrongly that all such numbers are prime. In fact, only the first five ($n = 0, 1, 2, 3$ and 4) are known to be prime; others have been shown to be composite. It is still not known whether there are more than five Fermat primes.

n	0	1	2	3	4
$2^{2^n} + 1$	3	5	17	257	65 537

What has all this got to do with Gauss's 17-gon? Gauss realised that the possibility of dividing a circle into a prime number of equal parts implied that the number in question must be a Fermat prime. He proved this by considering the solutions of certain algebraic equations; this is too advanced for us to pursue here. This meant that Gauss had settled the question of what whole number divisions of a circle are possible by Euclidean construction. Apart from Euclid's primes 2, 3 and 5, Gauss supplies three more: 17, 257 and 65 537.

If there are any other prime number divisions of the circle, they will be very large Fermat primes yet to be discovered and generally not thought to exist at all. This means that 7, 9, 11, 13, 14, 19, 21, 23, ... are truly impossible and not merely unknown Euclidean constructions. Of course, although it is theoretically possible to construct a 257-gon, and even a 65 537-gon, it is not practicable!

Reading

Ball, 1949; Coxeter, 1961; Hardy and Wright, 1938; Stewart, 1977.

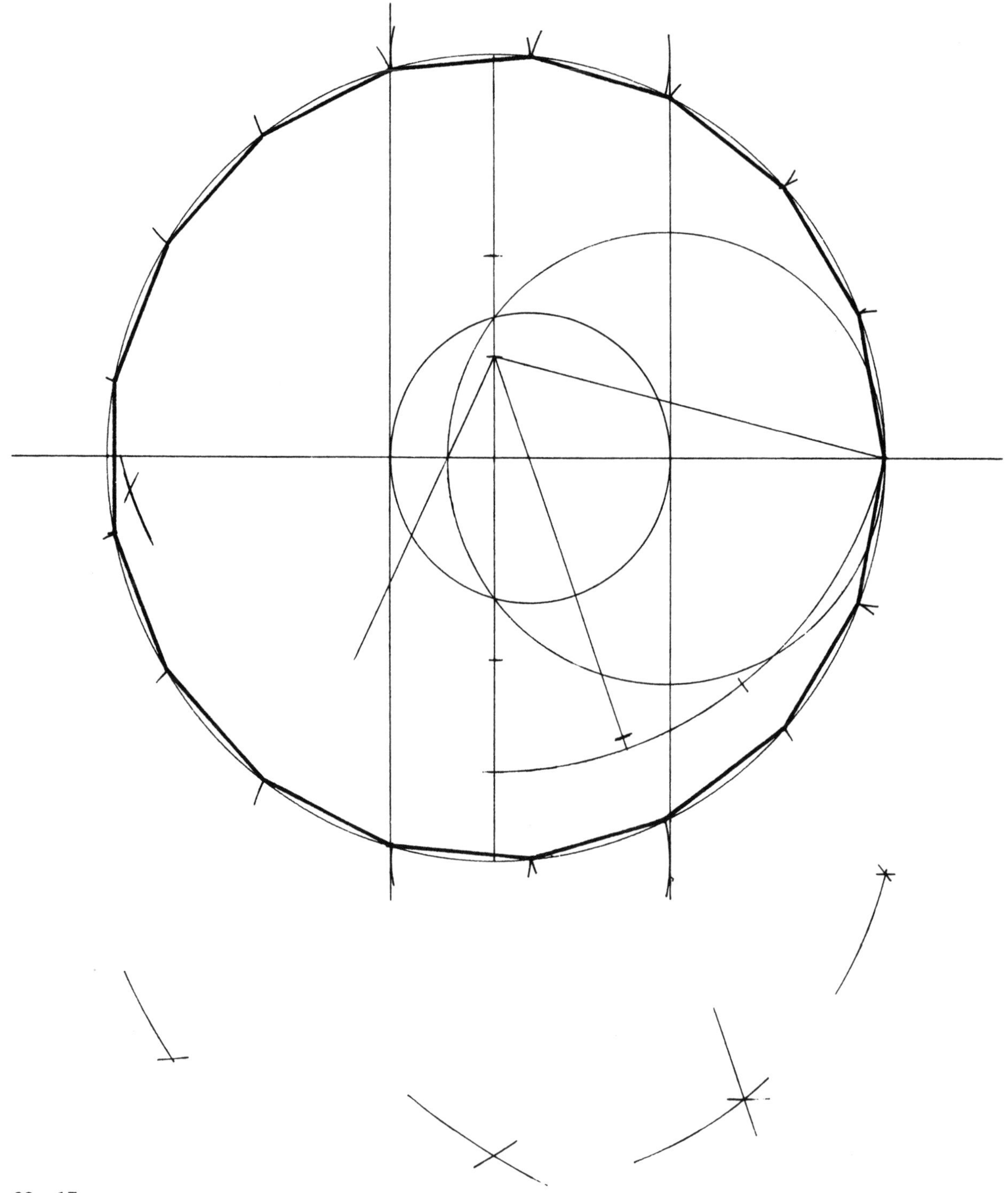

28 17-gon

1.5 The eight centres of a triangle

The drawings in this section provide simple exercises in the use of some of the Euclidean constructions introduced in Section 1.2, namely line bisection, angle bisection, erecting perpendiculars and dropping perpendiculars.

What we shall be doing is to identify and to construct each of the eight different centres associated with any given triangle. You should choose one fairly non-symmetrical triangle to work with and, having made several exact copies of this same triangle by tracing its vertices, start by completing each construction as a separate drawing. You can then go on to construct several of the centres together in a single drawing, so as to explore their interrelations. Finally, repeat these exercises with a variety of differently shaped triangles, including obtuse-angled triangles as well as acute-angled triangles. You should also try some of the special triangles, such as equilateral, isosceles and right-angled triangles.

CENTRE 1: THE CIRCUMCENTRE O

This is the centre of the *circumcircle*, which passes through the three vertices A, B and C of the given triangle (Figure 1.5A). To find the *circumcentre* construct the perpendicular bisectors of each of the three sides of triangle ABC. As the sides AB, BC and CA are chords of the circumcircle, their perpendicular bisectors must meet at the circumcentre.

Label the circumcentre O and the three mid-points of sides BC, CA and AB as X, Y and Z.

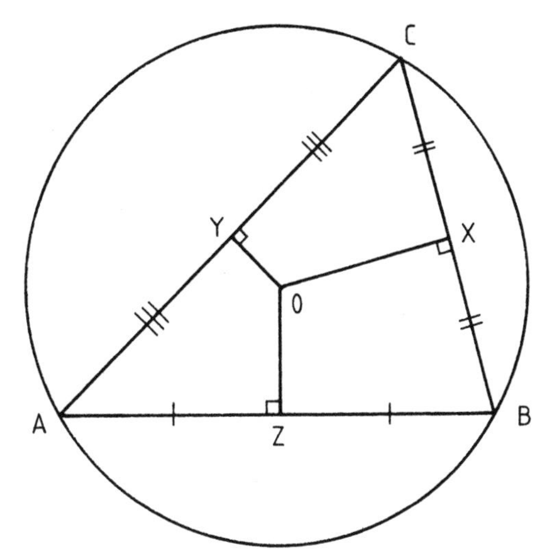

Figure 1.5A

CENTRE 2: THE CENTROID G

Each of the three lines AX, BY and CZ, connecting the vertex of a triangle to the mid-point of the opposite side is called a *median* (Figure 1.5B). The three medians meet at G, the *centroid*.

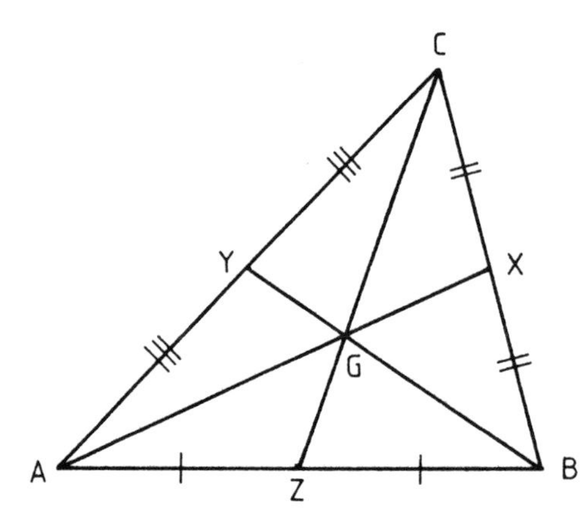

Figure 1.5B

If the triangle ABC were made of some material such as sheet metal, of uniform thickness and density, G would be its centre of gravity. It is fairly easy to see that each median bisects not only its corresponding side of the triangle but also all lines contained by the triangle which are parallel to that side (Figure 1.5C). This means that each median would provide a knife-edge line of balance for the triangle.

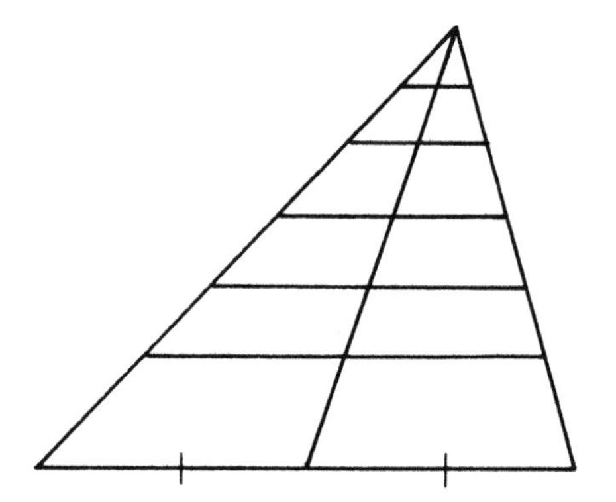

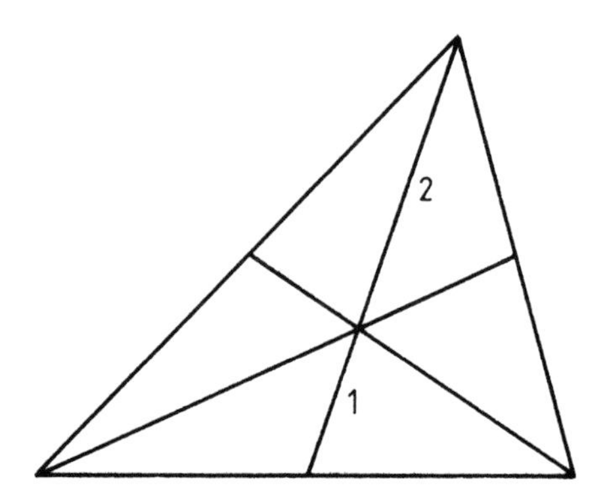

Figure 1.5C

The centroid G lies exactly a third of the way along each median from the side mid-point to the triangle vertex, as shown. The proof of this pleasing fact makes a nice exercise, whether you use similar triangles, vectors or integral calculus; the details are left for you to work out. Check by measurement.

By joining X, Y and Z, the mid-points of our given triangle ABC, you form the *auxiliary triangle* XYZ of triangle ABC (Figure 1.5D). It is easy to see via similar triangles that triangle XYZ is an exactly half-scale version of triangle ABC. So are triangles AZY, ZBX and YXC.

The medians of triangle ABC bisect the sides of the auxiliary triangle

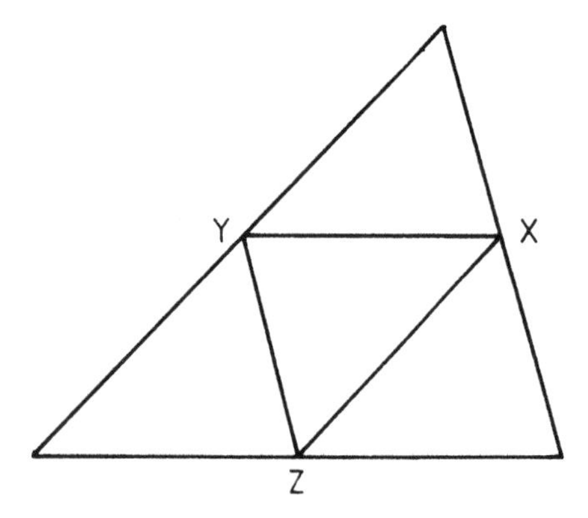

Figure 1.5D

XYZ, so that it is easy to construct the auxiliary triangle of the auxiliary triangle, and so on *ad infinitum*. These nested triangles converge rapidly on the point G, which is their common centroid.

CENTRE 3: THE ORTHOCENTRE H

To find this centre, construct the perpendicular from each vertex of triangle ABC to its opposite side (Figure 1.5E). The *orthocentre* H is the point where these three perpendiculars meet.

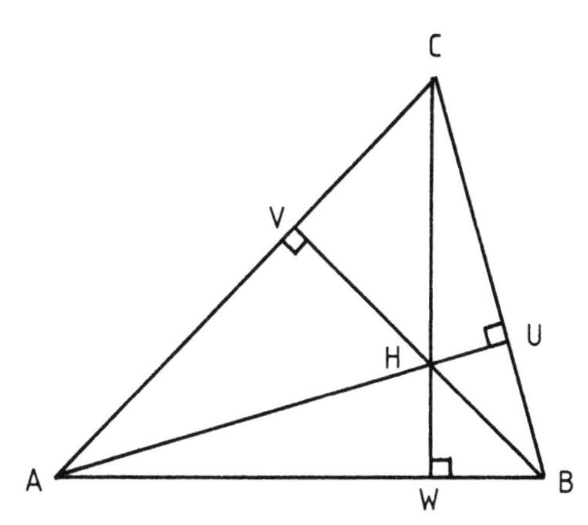

Figure 1.5E

Label the feet of the three perpendiculars U, V and W. Triangle UVW is called the altitude triangle (Figure 1.5F) and it has an interesting minimal property. It is the shortest closed path which touches all three sides of our given triangle ABC. If triangle ABC were made of perfect mirrors, UVWUVWUV... would be the path of a light beam destined to follow this path forever.

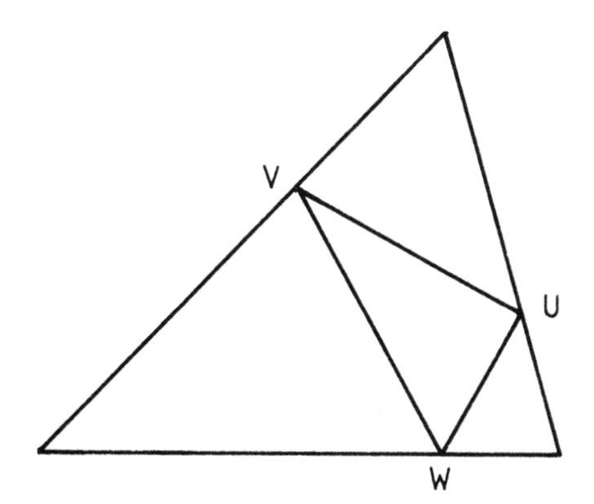

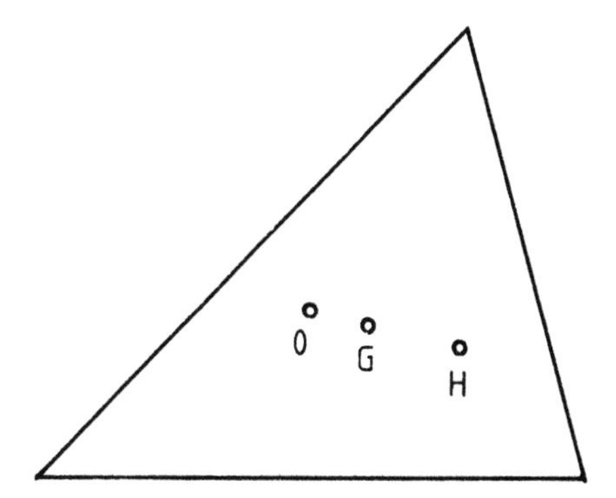

Figure 1.5F

If you construct the circumcentre O, the centroid G and the orthocentre H in the same triangle you will find that OGH forms a straight line, called the *Euler line* of the triangle ABC and that OG:GH = 1:2.

CENTRE 4: THE NINE-POINT CENTRE N

This is the centre of the *nine-point circle* of any triangle ABC, which passes through the following nine points associated with the triangle:

X, Y and Z (the vertices of the auxiliary triangle)
U, V and W (the vertices of the altitude triangle)
R, S and T (the mid-points of HA, HB and HC)

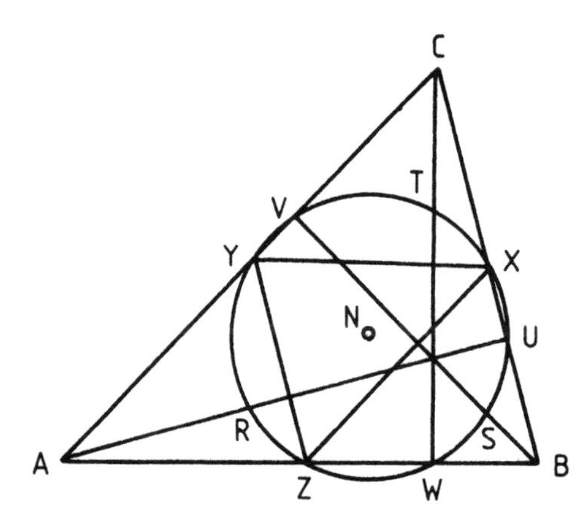

Figure 1.5G

It is quite easy to find the *nine-point centre* N, because it is the circumcentre of the auxiliary triangle XYZ (Figure 1.5G).

The nine-point centre N is also on the Euler line, with HG:GO:NG = 4:2:1. Note also that the circumcentre O of the triangle ABC is the orthocentre of the auxiliary triangle XYZ (Figure 1.5H).

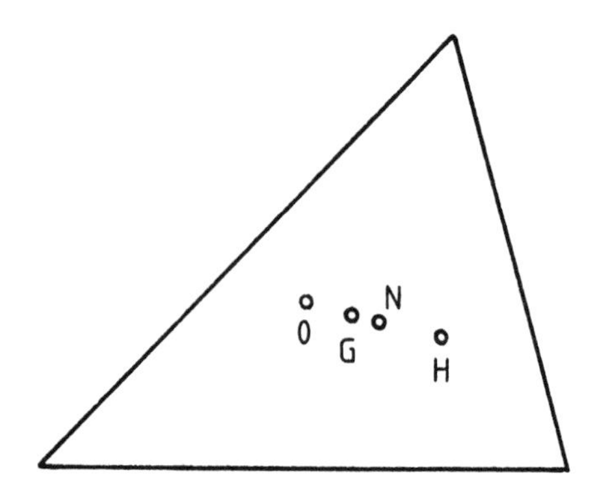

Figure 1.5H

CENTRE 5: THE INCENTRE I

The *incentre* of triangle ABC is the centre of the *incircle*, which touches all three sides of triangle ABC (Figure 1.5I). To find the incentre, construct the bisectors of angles A, B and C.

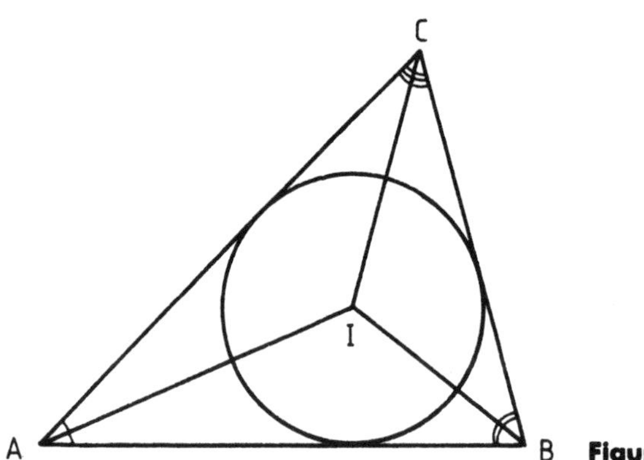

Figure 1.5I

CENTRES 6, 7 AND 8: THE EXCENTRES D, E AND F

These are the centres of the *excircles*, which each touch one of the sides of triangle ABC and the other two sides extended, as shown. Like the incentre, the *excentres* are found by bisecting the angles of triangle ABC, with this difference: bisect one internal angle and two external angles of the given triangle for each excentre (Drawing **29**). The bisector of the external angle and the bisector of the internal angle at any given vertex of a triangle are perpendicular to each other.

To construct the excircles of a triangle properly, you should work on a larger than usual scale, and you may need to use an extension bar on your regular compass, or else to use a beam compass to achieve large radii.

Note that I is the orthocentre of triangle DEF.

FEUERBACH'S THEOREM

The nine-point circle of a triangle ABC touches the incircle and each of the excircles.

You can verify this and all the other facts so far mentioned by construction.

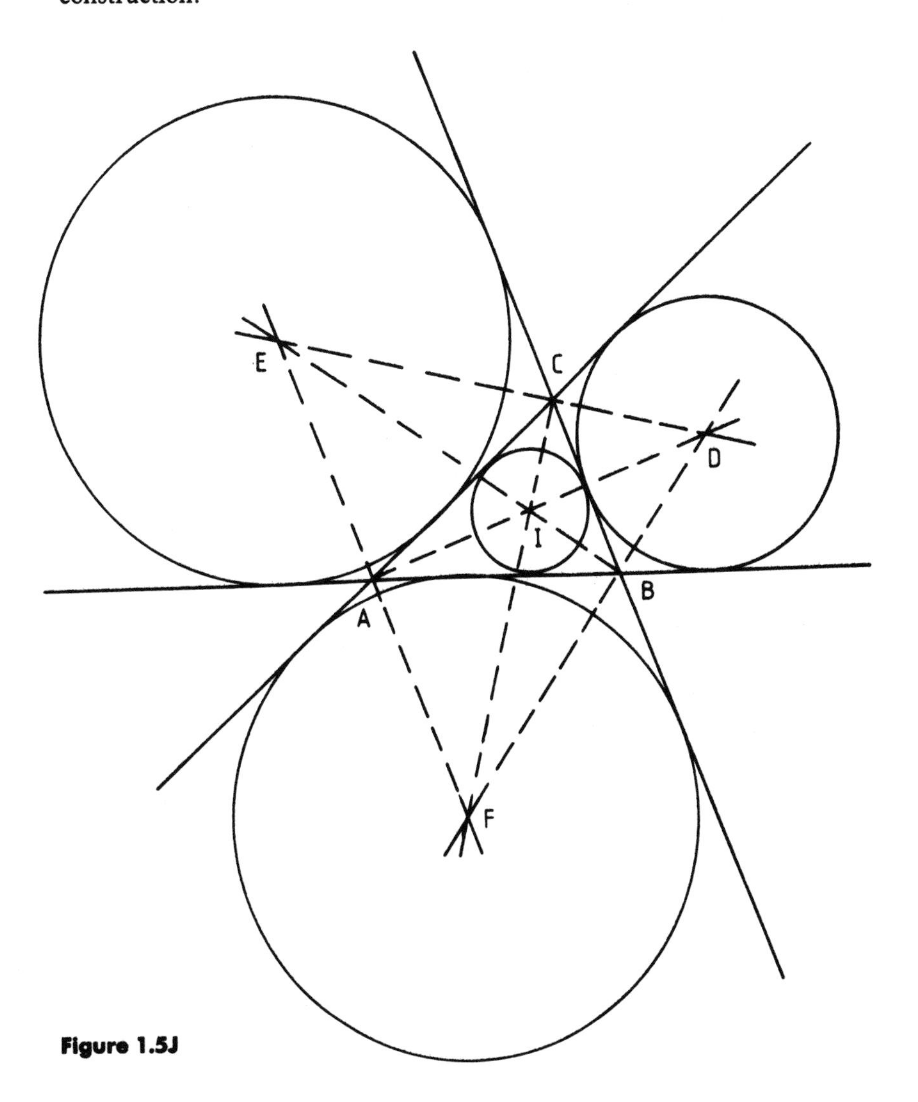

Figure 1.5J

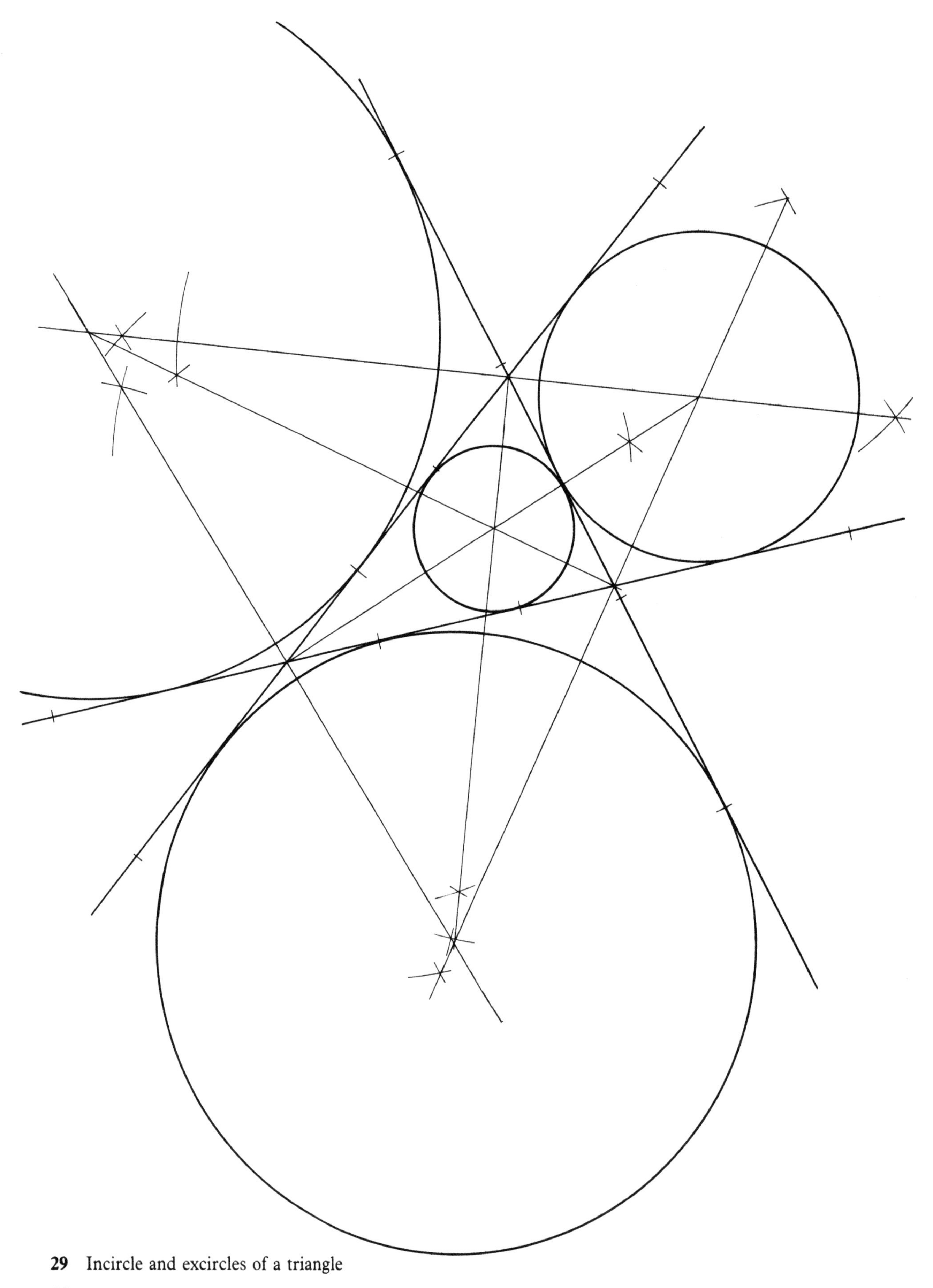

29 Incircle and excircles of a triangle

Exercise 3

1 Draw an acute triangle whose sides are all different, and then make careful copies of it to construct each of the following separately.

a The circumcentre O and the circumcircle.
b The centroid G.
c The orthocentre H.
d O, G and H together. Do they form a line? Is OG:GH = 1:2?
e The nine-point centre N, together with the nine points X, Y, Z, U, V, W, R, S and T.
f The incentre I and the incircle.
g The excentres D, E and F and the excircles.
h Incircle, excircles and the nine-point circles together. Is Feuerbach's theorem confirmed in your drawing?

2 Repeat the above exercises using an obtuse triangle.

Reading

Abbott, 1948; Coxeter, 1961; Pedoe, 1979; Roth, 1948.

1.6 Inverse points and mid-circles

INVERSION

This section introduces some of the compass constructions dealing with the transformation called *inversion*, which has the effect of turning the plane inside out about a given circle. The principle was invented by Jacob Steiner (1796–1863) and it led to a whole new branch of modern mathematics; yet the essential idea comes straight out of Euclid's *The Elements*, Vol. III. For that reason, the inverse constructions provide a natural part of Euclidean constructions which it so happens that Euclid never thought of.

The ideas in this section also serve to prepare us for later use of computer graphics to perform inversions, when we shall need to know the principle as a numerical rule. However, it is possible to do the compass constructions without understanding this. So this chapter is both a stepping stone to later chapters where algebra and computer graphics are involved as well as a compass exercise in its own right. Moreover, you should soon learn that the relations of inverse points and mid-circles (to be introduced shortly) are clearly visible. That is to say, the correctness of a drawing showing inverse points or a mid-circle is almost as directly apparent as a drawing of, for example, mirror symmetry or scaling similarity. Inverse relations have a sense of poise which is quite precise.

INVERSE POINTS

Figure 1.6A shows the principle of *inverse points*. Q is the inverse of P, and P is the inverse of Q with respect to the circle centred at O and the radius *r*.

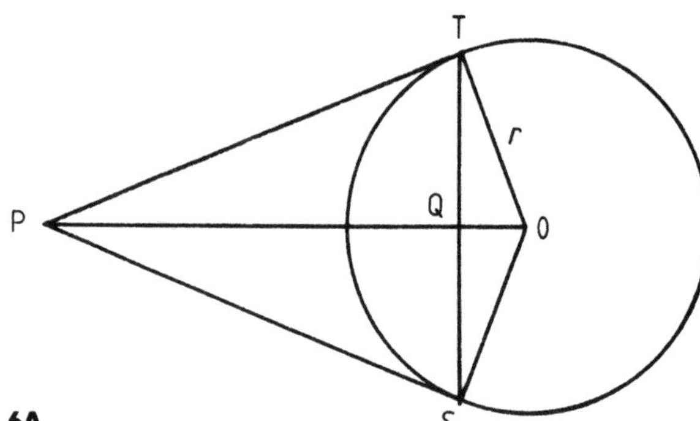

Figure 1.6A

For any point P external to the circle there is a pair of tangents which touch the circle at T and S. Q is the mid-point of TS.

Conversely, for any point Q internal to the circle, there is a chord perpendicular to OQ whose end points T and S have tangents to the circle which meet at P.

It is easy to see that every point has its own unique inverse with respect to any given circle. Only the centre of the circle in question seems to pose a problem. The nearer Q becomes to O, the centre, the further P becomes, without limit. So of O itself we say that its inverse is 'the point at infinity', and vice versa. Also note how, the nearer a point is to the circumference, the nearer will be its inverse to the circumference from the other side. So, in the limit, any point actually on the circumference is its own inverse with respect to that circle.

The above geometric definition of inverse points leads quite easily to the following numerical relation (look for similar triangles in Figure 1.6A):

$$\frac{OP}{r} = \frac{r}{OQ}$$

where O, P and Q are in a line with each other. The relation can be rewritten as

$$OP \times OQ = r^2 \text{ or } OQ = \frac{r^2}{OP}$$

Taking r as the unit of distance, we can say that OP is the *reciprocal* of OQ, and vice versa.

INVERSION IN A CIRCLE

With this definition of inverse points with respect to a given circle, we can define the transformation of *inversion*. To perform an inversion with respect to a given circle, move all points in the plane to the location of their inverses. Only the points on the circumference of the circle of inversion stay put, while all other points swap places in pairs.

Figures 1.6B, 1.6C and 1.6D illustrate the principle of inversion and the result of inverting a chessboard about the centre of the board. Note how the previously straight lines have become circles, except for the pair of lines which pass through the centre of inversion. The four-petalled inner shape is the inverse of the whole of the plane outside the chessboard, while the four inner squares of the chessboard now stretch to infinity.

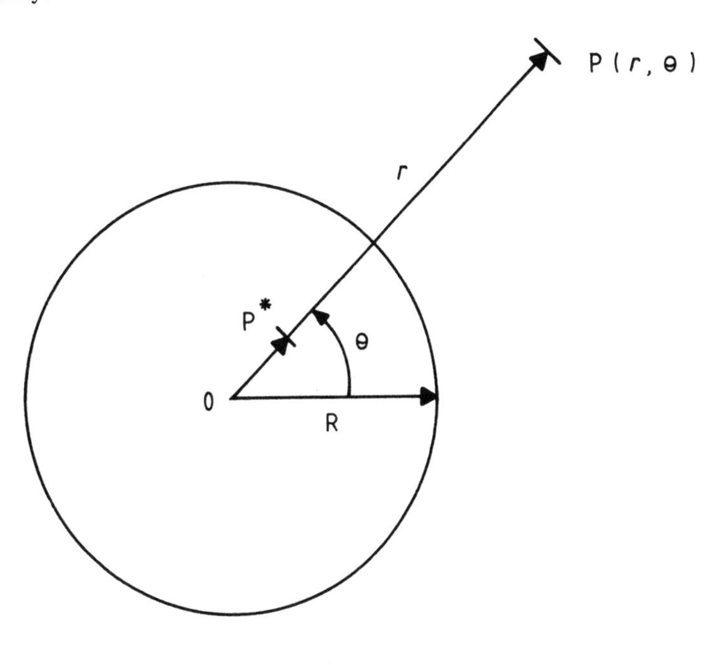

Figure 1.6B

Figure 1.6C

Figure 1.6D

Figure 1.6E

Incidentally, inversions with the same centre but different radii have exactly the same effect, except for an effect of scaling up or scaling down.

We shall be interested to know the effects of inversion on various curves and lines. Clearly a great deal of distortion of shape must take place if the whole of the outside of a circle is to be squeezed into the inside of a circle, and vice versa. Nevertheless, some important properties are preserved. To begin with, there is no cutting or crumpling of the plane, only stretching and compressing; although all other curves undergo a change in shape, the remarkable fact about inversion is that circles invert into circles, except those circles which pass through the centre of inversion whose inverses are straight lines. Conversely, lines invert into circles, except those passing through the centre of inversion, whose inverses are the same lines.

Figure 1.6E is something of a puzzle picture which is a summary of the effects of inversion on various lines and circles. Try to find the inverse image of each line and circle.

Note that if we regard a straight line as a circle of infinite radius, we are able to say that inversion preserves circles without exception. Another important property, whose proof will be left until later, is that inversion preserves angles.

ORTHOGONAL CIRCLES

Two circles are said to be *orthogonal* if they cut each other at right angles (Figure 1.6F). This means that at the point of intersection of a pair of orthogonal circles the radius of one circle is a tangent to the other circle, and vice versa (Figure 1.6G). It is this fact which points the way to constructing a circle about a centre P which is orthogonal to a given circle centred at O. Take the tangent PT as the radius of the required circle.

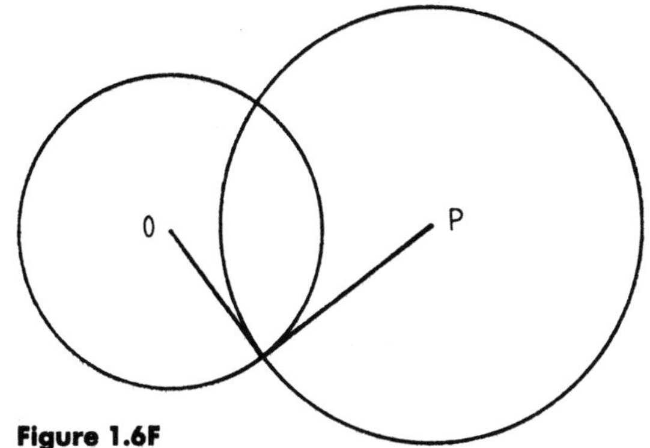

Figure 1.6F

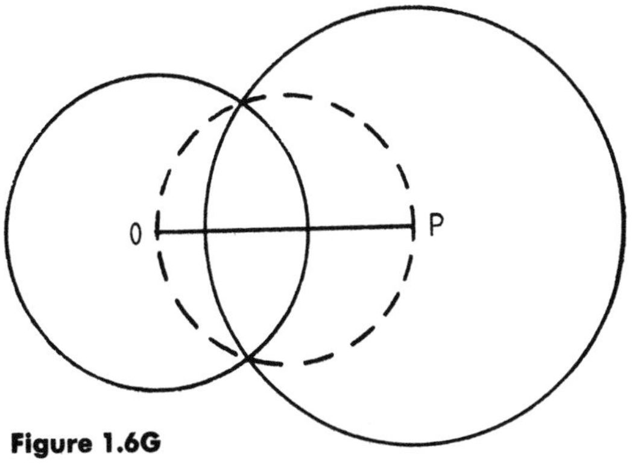

O P

Figure 1.6G

There is a well-known theorem in Euclid's *The Elements*, Vol. III, p. 36, dealing with a tangent OT to a given circle and any secant OPQ which states that

$$OP \times OQ = OT^2$$

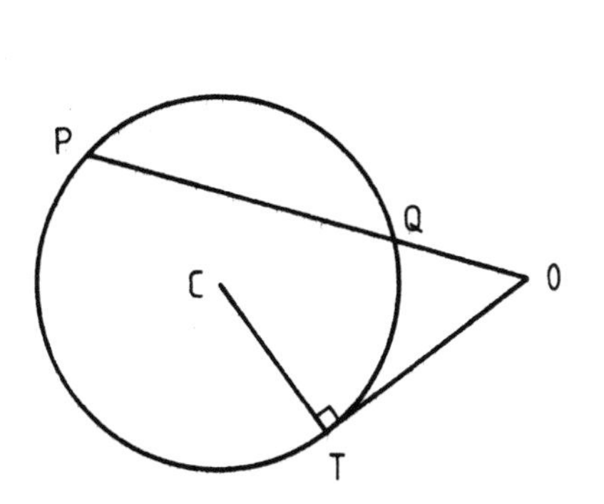

Figure 1.6H

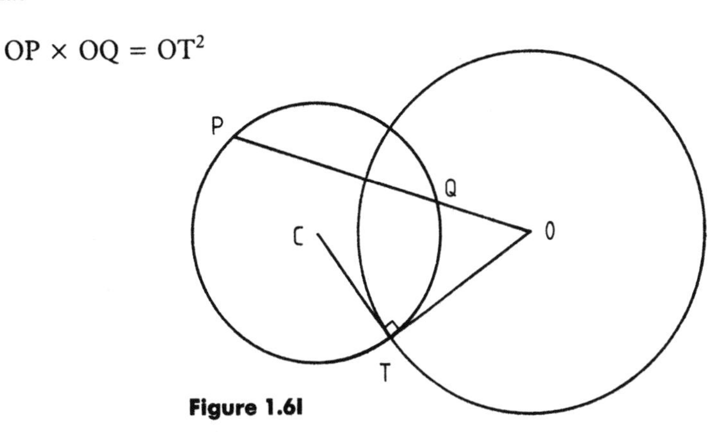

Figure 1.6I

It is easy to see from this formula that P and Q must be each other's inverse with respect to a circle centred at O and of radius OT. Then, by comparing Figures 1.6H and 1.6I, it becomes clear that any circle orthogonal to a circle of inversion must invert into itself. All the points in the exterior arc swap places with points of the interior arc.

MID-CIRCLES

A *mid-circle* of two given circles is the circle which would invert each of the two given circles into the other.

There are five cases to consider (Figure 1.6J).

- **a** When the two given circles intersect each other.
- **b** When the two given circles touch each other externally.
- **c** When the two given circles are external to each other.
- **d** When there is one given circle inside another given circle.
- **e** When one given circle touches another given circle internally.

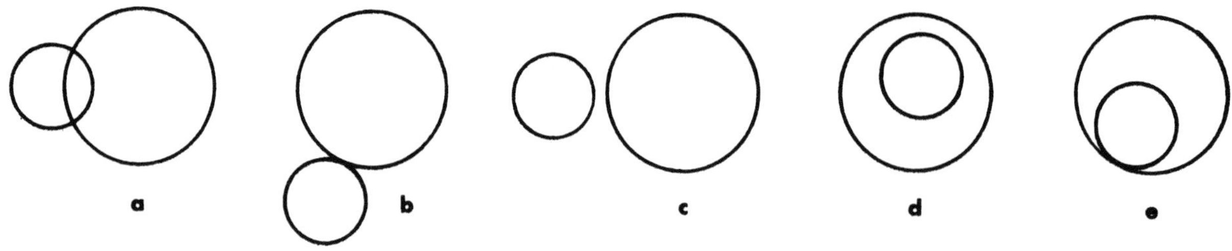

Figure 1.6J

The task of constructing the mid-circle in each of these cases sets us a variety of problems, ranging from the quite simple to the rather complicated. Cases **a** and **b** present no difficulty at all. As Figures 1.6K and 1.6L show, the two given circles have a pair of common tangents which meet at the centre of the required mid-circle. The radius of the mid-circle is obtained from the point(s) of contact (intersection) of the two given circles as shown.

Figure 1.6K **Figure 1.6L**

Actually, there is a second mid-circle for case **b**, the case of two intersecting circles, but its discovery and construction will be left as an exercise.

For case **c**, where the two given circles are external to each other, we again construct a pair of common tangents to find their point of intersection, which is the centre of the mid-circle. The radius of the

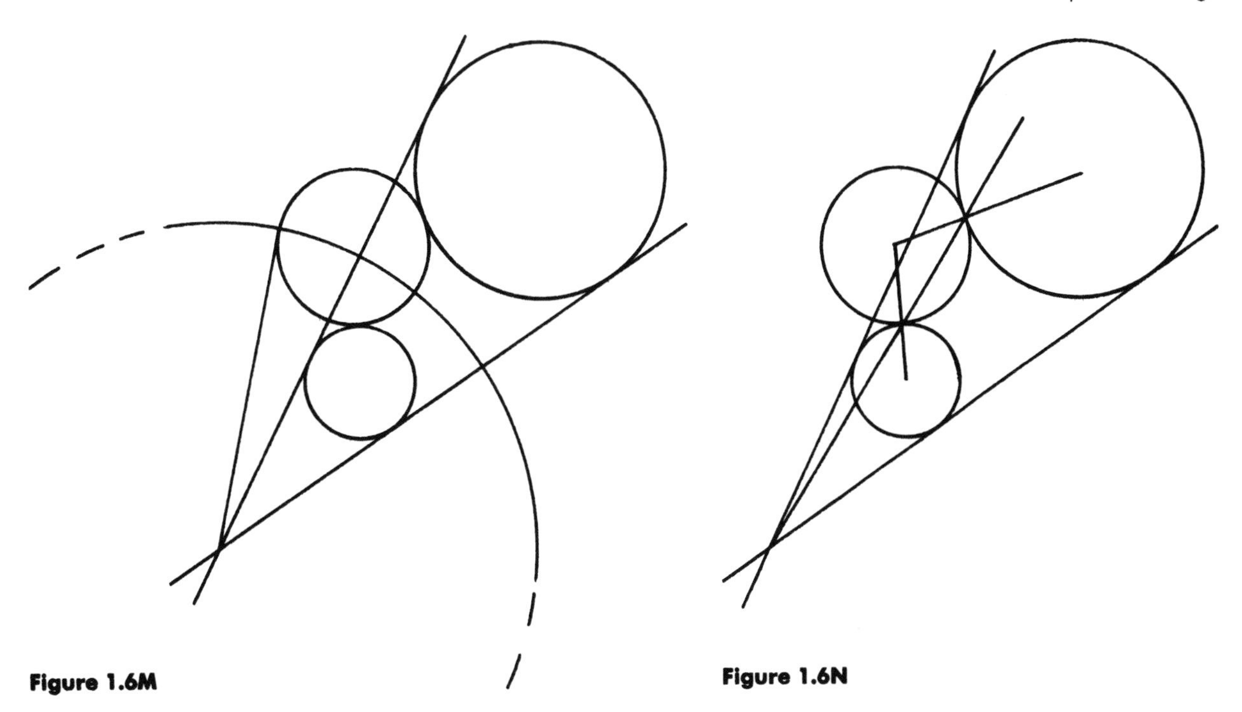

Figure 1.6M

Figure 1.6N

mid-circle, however, is a little more difficult to obtain. Construct any circle which is tangent to both of the given circles and then find the point of tangency on this circle from the centre of the mid-circle. This gives the required radius of the mid-circle, as shown in Figure 1.6M.

The explanation for the above construction is that the circle constructed to be tangent to both of the given circles must invert into itself with respect to the mid-circle and so must be orthogonal to the mid-circle.

Note that, although you may well find it fairly easy to construct a circle tangent to two given circles by careful trial and error, in the spirit of the trickier short cuts of Section 1.2, here are the details of the strictly possible Euclidean steps: draw a line from the centre of the mid-circle to cut the two given circles (Figure 1.6N) to provide the points of tangency and then find the centre of the tangent circle by extending the pair of radii at these points to meet at the required centre.

Case **d**, when one of the given circles is inside the other, presents the hardest mid-circle construction because we have no immediate way of finding the centre of the mid-circle. Proceed as follows. First, construct

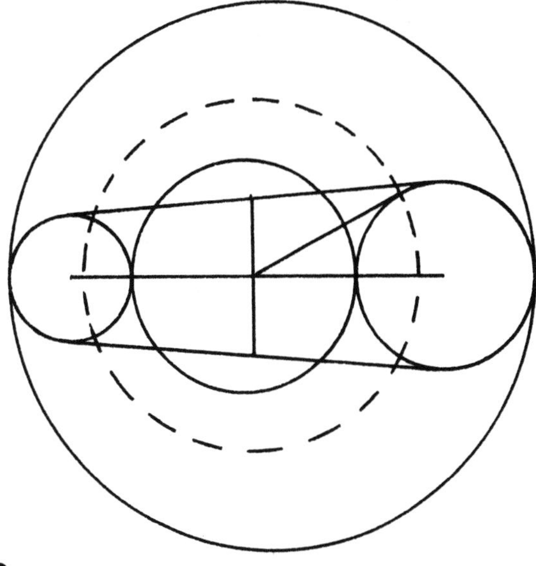

Figure 1.6O

the line through both centres of the given circles and, on this line, construct a pair of circles each of which is tangent to both of the given circles, as shown in Figure 1.6O. Their centres are found by line bisection. Next, construct a pair of common tangents to these constructed circles and find their mid-points. The line through these mid-points cuts the line of centres at the required centre of the mid-circle. To find its radius, construct a point of tangency from this centre to one of the constructed tangent circles.

Part of the explanation of the above construction is that the two constructed tangent circles must invert into themselves with respect to the mid-circle, and are thus orthogonal to it. However, a complete explanation here would take us too far into purely mathematical argument; instead, try to verify its validity by careful practice. If you wish to track down the full proof, you should find enough clues in the following sections.

Case **e**, where one of the given circles is inside and touching the other, presents a somewhat easier mid-circle construction. This is left for you to discover and try.

THE RADICAL AXIS

The *power* of a point P with respect to a circle of radius r and centre O is defined as $OP^2 - r^2$. It is easy to see that for points external to the circle the power is the square of the length of the tangent to the circle squared. For internal points the power is minus the square of the length of half the chord perpendicular to OP squared. (Incidentally, the power is also the product on any secant or chord; see Figure 1.6P.) The *radical axis* of two given circles is the path of points whose powers with respect to both circles are equal. In other words, the radical axis is the set of all points with equal tangents to the two given circles.

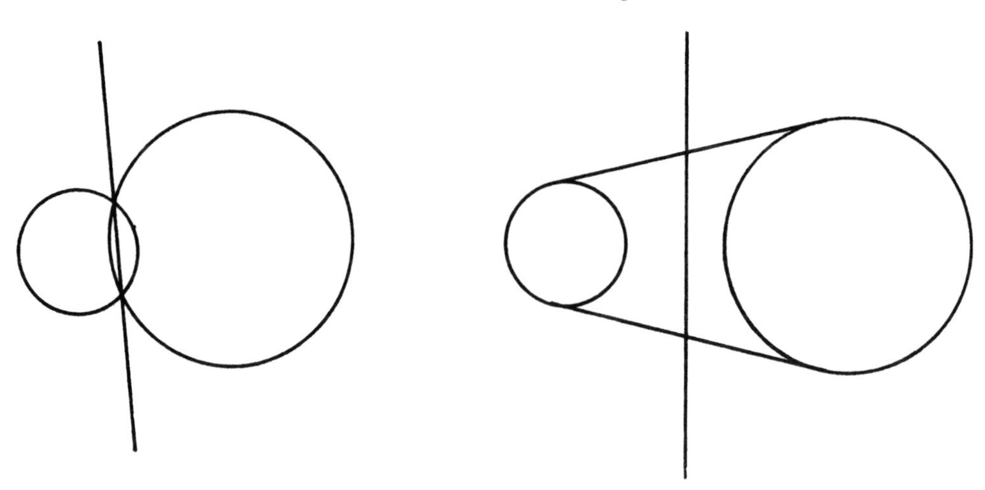

Figure 1.6P

It is quite easy to show that the radical axis of two given circles must be a straight line perpendicular to the line of the circle centres. This means that the radical axis of two intersecting circles is the line through the points of intersection. For two non-intersecting circles with a pair of common tangents, the radical axis bisects the tangents.

COAXAL CIRCLES

All circles which share a common radical axis with any given circle are *coaxal.* Their centres are collinear and either they all intersect at a common pair of intersection points, i.e. an *intersecting* set of coaxal circles (Drawing **30**), or else none of the circles intersect, i.e. a *non-intersecting* set

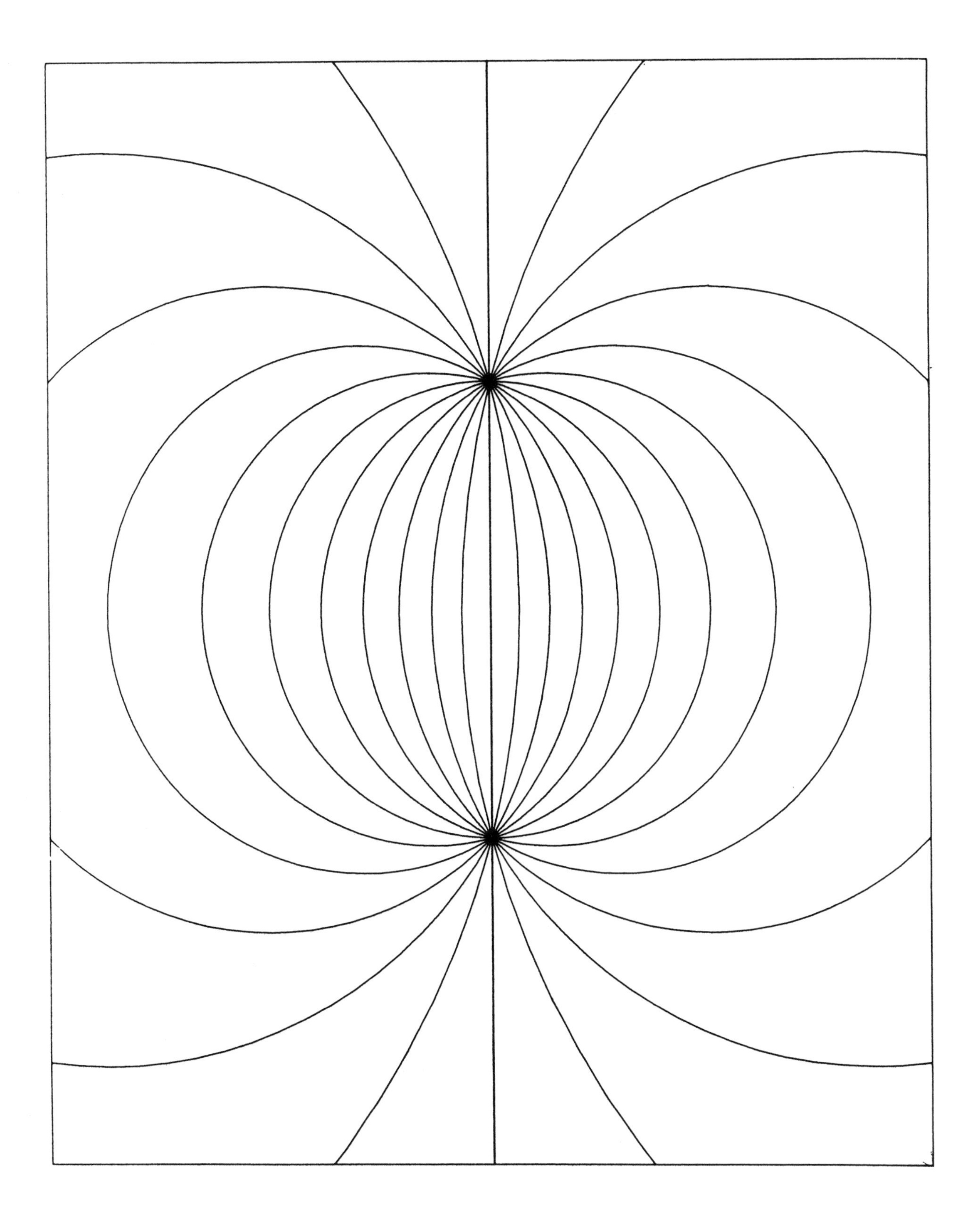

30 Intersecting set of coaxal circles

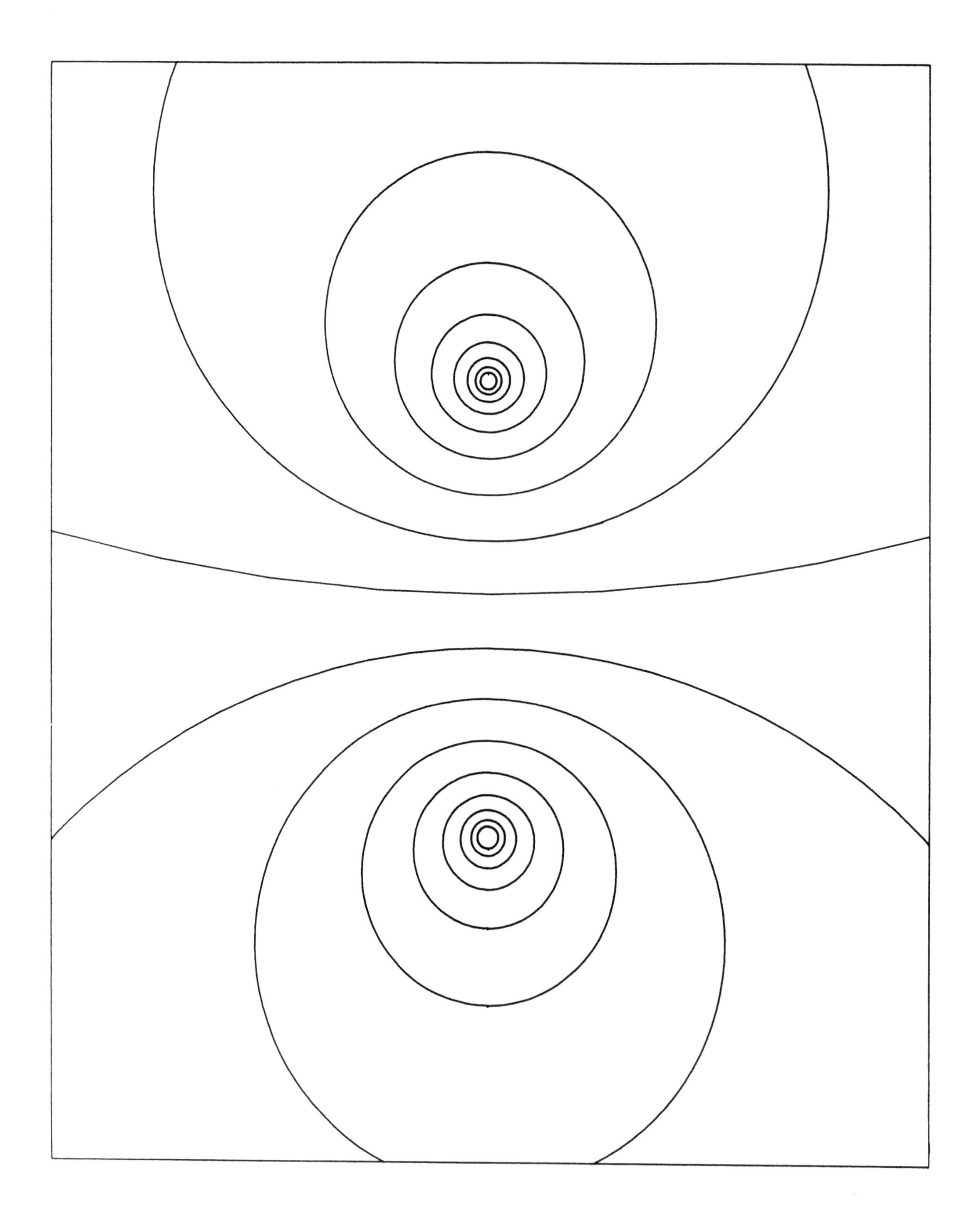

31 Non-intersecting set of coaxal circles

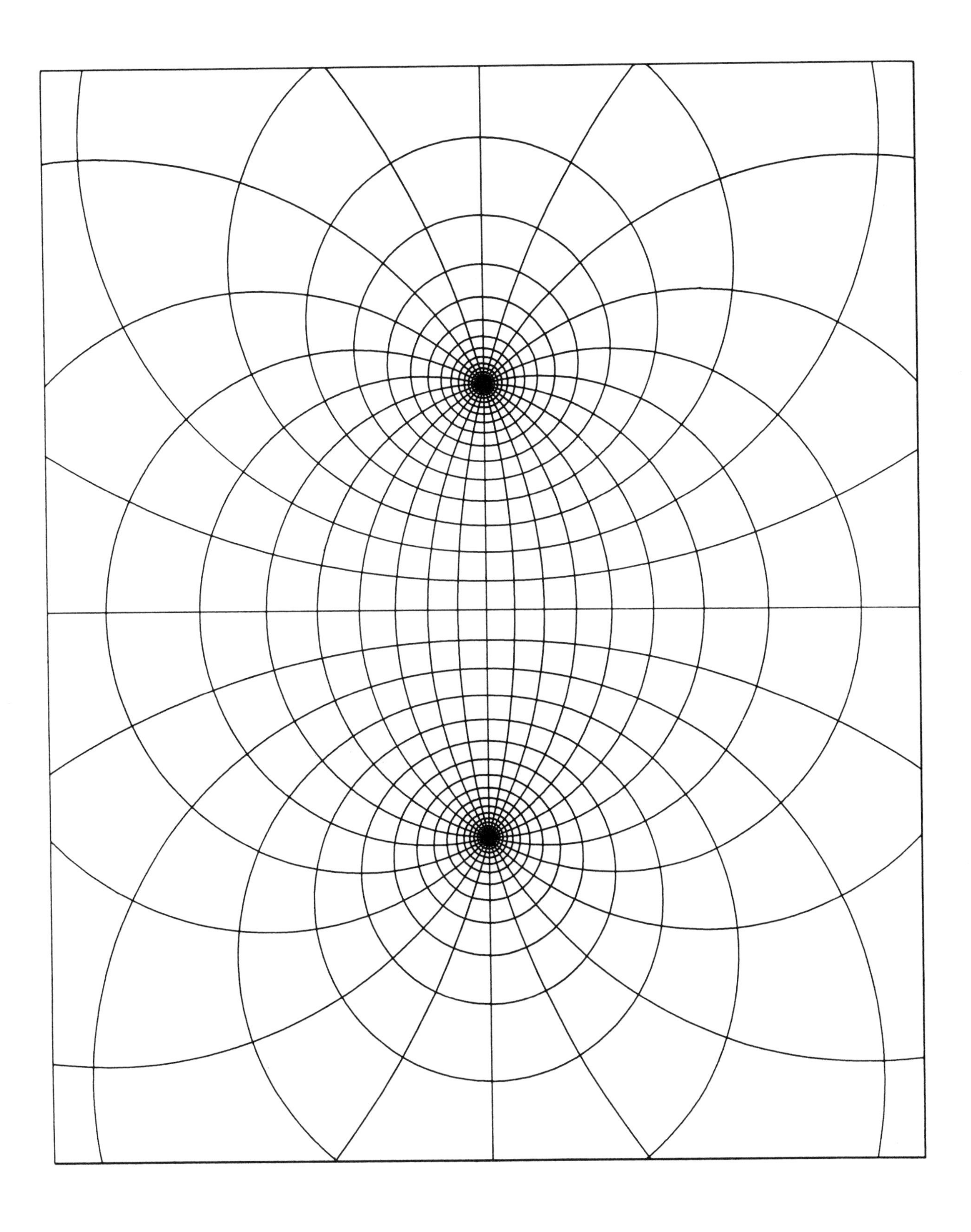

32 Orthogonal sets of coaxal circles

of coaxal circles (Drawing **31**). In each case the set includes the radical axis as a special case where the circle radius is infinite.

Drawing **32** shows two sets of coaxal circles together, one intersecting and the other non-intersecting. All the circles in one set are orthogonal to all the circles in the other set. The radical axis of each set is the line of centres of the other set. The two intersection points in the intersecting set are extreme members of the non-intersecting set and are called the *limiting points* of this set.

The subject of coaxal circles provides some excellent problems in analytical geometry which are outside the scope of this book. We shall just mention some of their properties before tackling our chosen construction problem, the limiting points of a pair of given circles. First of all, note that a pair of circles and their mid-circle are coaxal. Second, the path of all points whose powers with respect to two given circles are in a constant ratio is a circle coaxal with the two given circles. By changing this ratio, we can generate each of the circles in the coaxal set in turn. Third, the bipolar equations for a point P given by AP = kBP define a non-intersecting set of coaxal circles, each circle given by a different value of k, ranging from 0 to infinity. Finally, Drawing **32** is the inverted image of a set of concentric circles and a set of radial lines through their common centre.

To construct the limiting points of two given circles proceed as follows. Construct the line of their centres and their radical axis and, taking the intersection of these two lines as centre, construct a circle orthogonal to the given circles. The points at which this circle cuts the line of centres are the required limiting points (Figure 1.6Q).

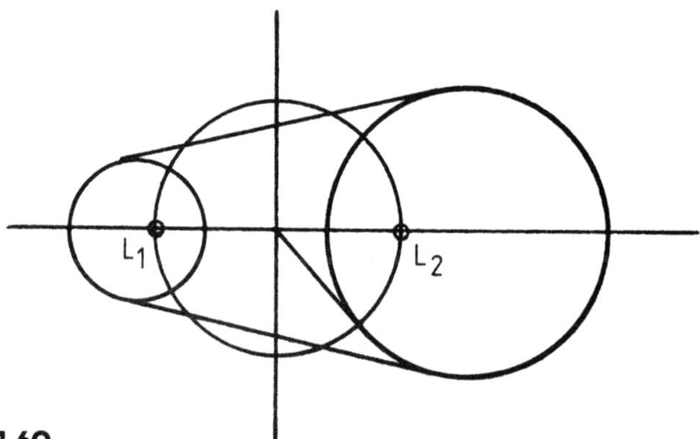

Figure 1.6Q

Exercise 4

1 Construct a circle and, choosing a point outside it, construct its inverse. Repeat this exercise for an internal point.

2 Construct two circles. Then construct the inverse of one of these circles in the other circle.

3 Construct a circle and then, choosing an external point as its centre, construct a circle orthogonal to the first circle.

4 Construct a circle and then, on its diameter suitably extended on either side, construct several circles orthogonal to the first. What do you notice?

5 Construct the mid-circle for each of the five cases of two circles considered in this section.

6 Construct the second mid-circle of two intersecting circles.

7 What is the mid-circle of the following?
a Two equal and intersecting circles.
b Two equal and non-intersecting circles.
c Two concentric circles with respective radii of 2 units and 8 units.

8 By constructing three circles, each of which intersects the other two, confirm that the three radical axes of three circles taken in pairs meet at a common point, called the *radical centre*. Note that the three circles should not have collinear centres.

9 Construct the radical axis of two circles, one of which is inside the other.

10 Construct the limiting points of two circles, one of which is inside the other.

11 Sketch the inverse of a hand (or handed shape, such as F) in a circle to confirm that inversion, like mirror reflection, reverses handedness.

Reading

Coxeter, 1961; Dixon, 1983a; Pedoe, 1979.

2 String Drawings

In Section 1.1 we saw how a smooth curve can be made up by joining together a number of circular arcs, each of a different radius. The aim of the present chapter is to complete that story by increasing the number of arcs, and therefore the centres of curvature, to the stage where we can imagine a continuously changing radius of curvature. Again, we shall use the curve of an egg shape to illustrate this, but the lesson applies to all smooth curves. That is to say, any curve may be thought of as infinitely many, infinitely short circular arcs joined together.

There are several ways of using string to construct mathematical curves, but we are here only concerned with one of them, the method called an *involute* construction.

THE INVOLUTES OF AN EVOLUTE

Let us again consider the compass eggs in Section 1.1, and in particular the five-centred egg. Just as it is possible to draw a circle using a stretched string instead of a compass, so it is possible to construct the egg by using the string and some pins (nails) in the following way (Drawing **33**). With the drawing paper attached to a board, insert pins firmly at each of the five centres, and tie the string to the centre of the arc whose radius is largest. By attaching a pen to some other point on the string, it is possible to draw the curve in one sweep. The string is maintained in a taut condition as it passes over each of the pins in turn (Figure 2.1A).

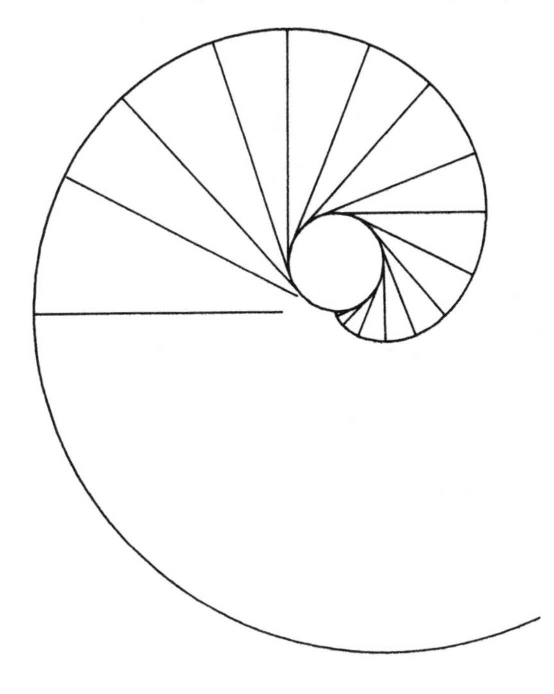

Figure 2.1A

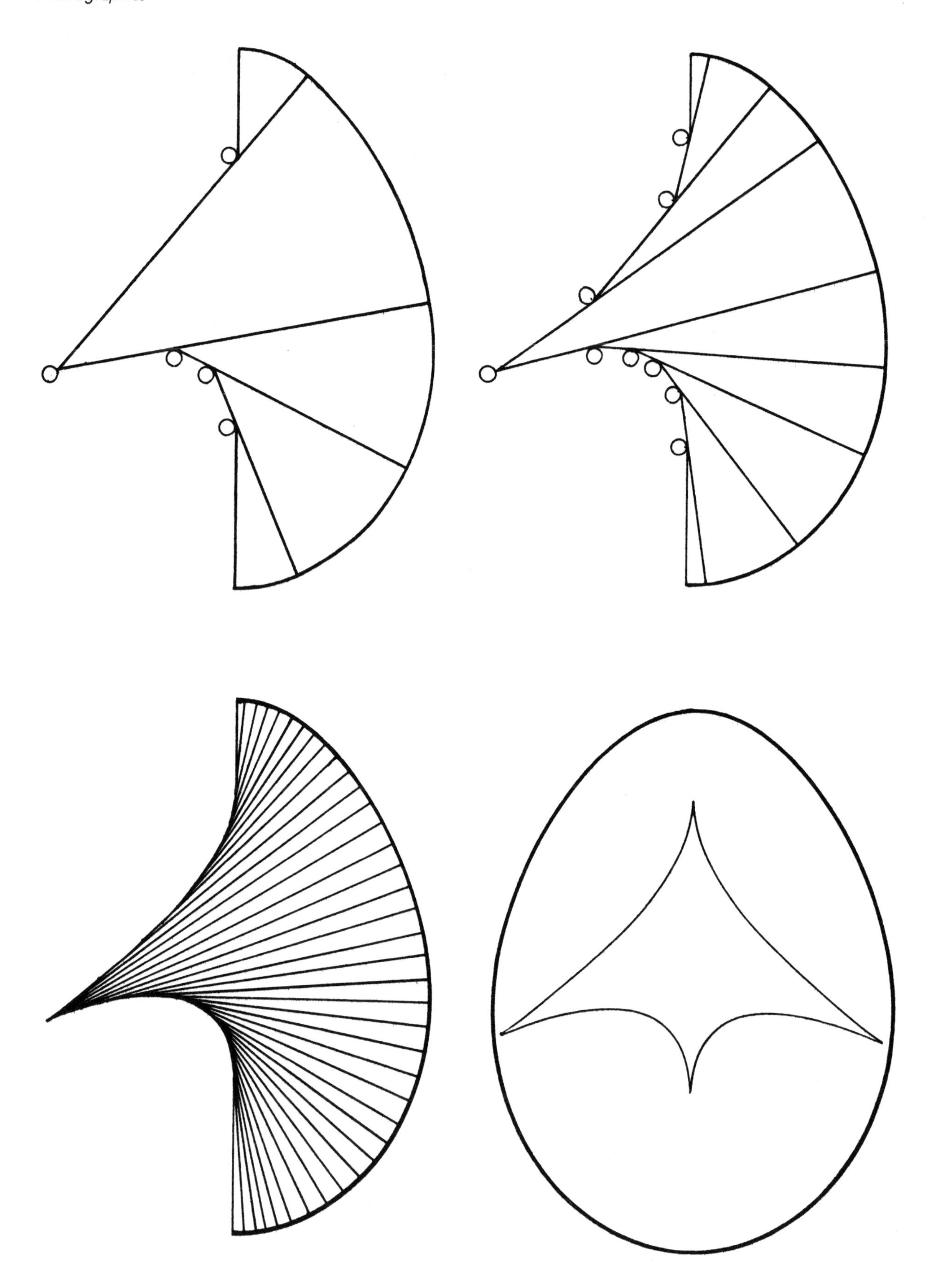

33 String-drawn egg

Having performed this construction with five centres, it is easy to put in more and more pins, as shown, so increasing the number of centres, and therefore the number of separate circular arcs. In the limit we can imagine infinitely many centres forming a smooth curve; this is called the *evolute* of the egg curve.

The curve traced out by a point on the stretched string wrapping and unwrapping over an evolute, like this, is called an *involute* curve. Clearly, for every evolute curve there are as many distinct involutes as there are distinct points on the string (i.e. infinitely many).

Try to perform such a construction and to discover that a curve with a continuously changing radius of curvature, or at least very many radii of curvature, is visibly different from one constructed from a few circular arcs.

PRACTICAL ADVICE

The test of accuracy in a string drawing is to try several sweeps of the pen backwards and forwards to see whether it traces out the same curve (involute). Make sure that the pins are firm and vertically fixed. Choose a good-quality thread or cord with as little stretchiness as possible. You will also need to give careful attention to the manner in which the pen is attached to the string. There is plenty of scope for sloppiness here. I find that a shirt button gives an ideal solution; attach the button to the string via one of its cotton holes and place the pen nib or pencil in another of its cotton holes. This has the effect of keeping the point of attachment very close to the paper surface but without the tendency for the pen to slip out — something which happens all too easily when a simple loop of the string is used.

One final suggestion which goes well with this exercise is to construct a template using stiff card. First, draw the curve on the card as described above. Cut along the curve using a sharp knife or scalpel. Then gently smoothe out the template edge using fine sandpaper. Any slight corner or irregularity is soon removed like this. If you intend to trace round the template with a wet ink pen, it will be necessary to chamfer the lower edge just enough to prevent smudging due to capillary effects. The curve traced out using your template will be found to be satisfyingly smooth and easy to draw.

OTHER EVOLUTES

Every smoothly flowing curve has its own evolute, although it is not usually very easy to find it accurately. Mathematicians tackle this problem by means of advanced calculus. However, having seen the evolute of an egg, as illustrated above, you should be able to imagine what, for example, the evolute of an ellipse should look like. Study various well-formed curves and attempt to construct their evolutes approximately.

This can be done by using the property that the evolute is the *envelope* of normals to the curve (involute). This property is illustrated in Figure 2.1A, because the radius of curvature at any point on the involute curve is also a normal to the curve.

Working in the opposite direction, experiment with various evolutes to see what involutes they give. Set up various arrangements of pins and see what happens.

One evolute which is easy to try is the circle. (It was first used long ago for drawing spirals.) Use a pencil or some other circular cylinder set up vertically on the drawing board. Attach the string to the cylinder and allow the pen to move so that the string wraps round or unwraps from the cylinder. The involutes of this construction approximate very well to Archimedes' spiral (Figure 2.1B) which is defined as a spiral which increases its radius by equal amounts in equal degrees of turning.

Exercise 5

1 By trying out various shapes of evolute, try to obtain various shapes of eggs.

2 Get hold of a well-defined curve, from decorative art, for example, or a mathematical illlustration of an ellipse and, by fitting either circles or normals to the curve, try to obtain an approximate evolute. You can check your model by drawing an involute to compare with the curve example.

3 The measure of *curvature* of a curve at a given point where the radius of curvature is r, is given by $c = 1/r$.

- **a** What is the curvature of a straight line?
- **b** Which has the greatest curvature, a small circle or a large circle?
- **c** For a smooth flowing curve, is there a single measure of curvature for every point on the curve?

Reading

Courant, 1934; Cundy and Rollett, 1961; Lamb, 1897; Lockwood, 1961; Thompson, 1917.

3 Perspective Drawings

The ideas and practice of perspective drawing are due to Filippo Brunelleschi (1377–1446), the great Florentine architect who also started the classical revival in architecture. The effects on picture making were immediate and dramatic. For the first time, artists confidently depicted three-dimensional space on the picture plane and, in particular, they learnt the rules for depicting buildings and other objects made up of straight lines. The laws of perspective are essentially about how to depict lines and parallels.

Brunelleschi's first idea was as follows. Imagine that you are standing before the scene to be painted, looking through one eye, and then suppose that a picture plane is set between you and the scene. Next, and without moving your head or the picture plane, imagine a *ray* from your fixed point of view tracing out the scene. The image of each point in the scene is the point at which its ray intersects the picture plane (Figure 3.1A).

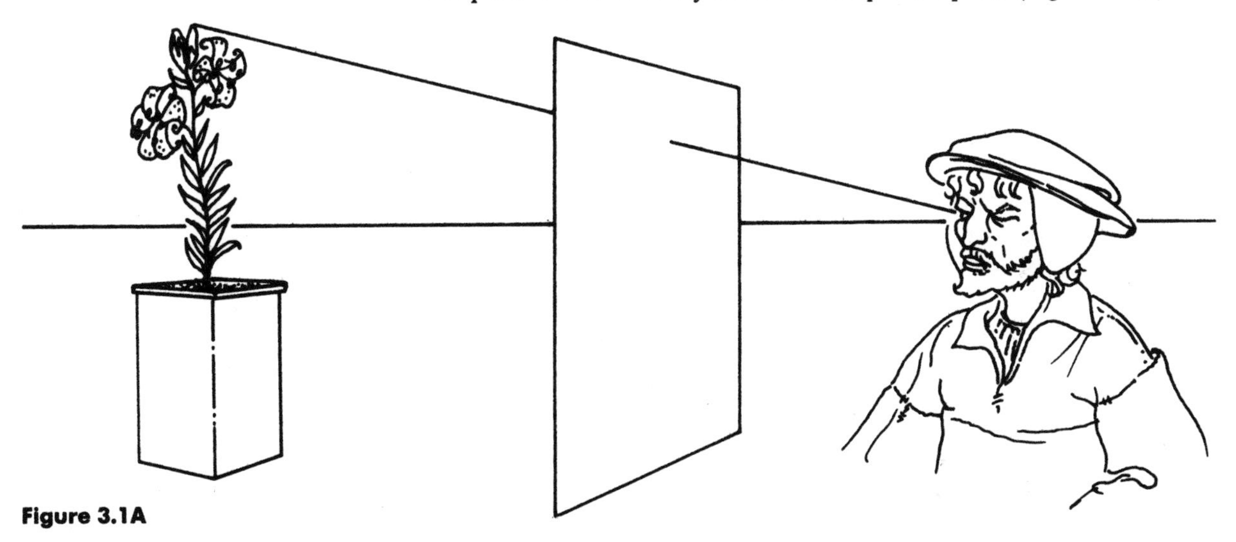

Figure 3.1A

Brunelleschi's second idea was to deduce from this principle of point projection the rules which must apply to the images of straight lines and sets of parallel lines. So, instead of actually performing such a projection of a scene, he was able to *construct* a perspective drawing. Although artists experimented with various devices for performing a projection, no easy method for doing this was found until the invention of photography 400 years later. By contrast, the rules for constructing perspective pictures of straight-lined objects turned out to be simple and powerful.

It is these rules (the horizon appears as a line, parallels appear to meet at vanishing points etc.) that we shall look at in this chapter. We tackle some of the key problems to be solved, such as the location of vanishing points and the rules of foreshortening.

To illustrate this, I shall be drawing rectangular grids, endlessly straight railway tracks and box-like houses. Once you have learnt the handful or so of rules, you will find the scope for variety and levels of detail to be endless, but this will be left for you to explore. It is even possible to apply the rules to such tricky subjects as a 'spiral' staircase.

HORIZON LINE AND VANISHING POINTS

Theorem 1 The horizon appears as a line.

Imagine that you are standing on an endless plane, rather like an extremely flat desert, or out at sea. The rays from your point of view which reach out to points infinitely far away in all directions of the plane form a second plane, parallel to the one on which you are standing. This second plane cuts the picture plane in a line.

Theorem 2 Straight lines in the scene appear straight in the picture.
Proof Rays from your point of view to a line describe a plane, which therefore cuts the picture in a line.

Theorem 3 Sets of parallel lines meet at a *vanishing point*.
Proof Imagine that from your point of view you attempt to look further and further into the distance along any one of the parallel lines in question. Eventually you will be looking in a direction parallel to the line. Where this ray cuts the picture plane is the vanishing point for all lines parallel to that direction.

Corollary Different sets of parallels have different vanishing points.

Where in relation to each other do we place the various vanishing points? We shall tackle this problem step by step.

ONE-POINT, TWO-POINT AND THREE-POINT PERSPECTIVES

A box-like building has three sets of parallel lines, each perpendicular to the other two. One set of lines is vertical, and the other two sets are horizontal, pointing, for example, north–south and east–west respectively. We can greatly simplify the construction problem if we choose to show one or two of the three parallel sets as parallel to the picture plane.

Theorem 4 Lines parallel to the picture plane appear parallel and therefore have no vanishing point.
Proof The ray which attempts to trace such lines further and further into the distance eventually becomes parallel to the picture plane and so has no vanishing point. Lines in a plane which do not meet are parallel.

A *one-point perspective drawing* (Figure 3.1B) of our box is achieved by setting two of the parallel sets parallel to the picture and drawing them as parallel. Only one vanishing point, therefore, is needed.

In a *two-point perspective drawing* (Figure 3.1B), only one set of the parallels remains parallel to the picture plane, usually the vertical lines. We now need to locate two vanishing points. Of course, if the box-like building is horizontal (not tilting or warped), the vanishing points for the 'east–west' and 'north–south' lines must both be on the horizon. You are free to place the first vanishing point anywhere on the horizon line but, having done so, you are restricted in choice for the second vanishing point.

What guidelines are there for this problem? Well, imagine once more

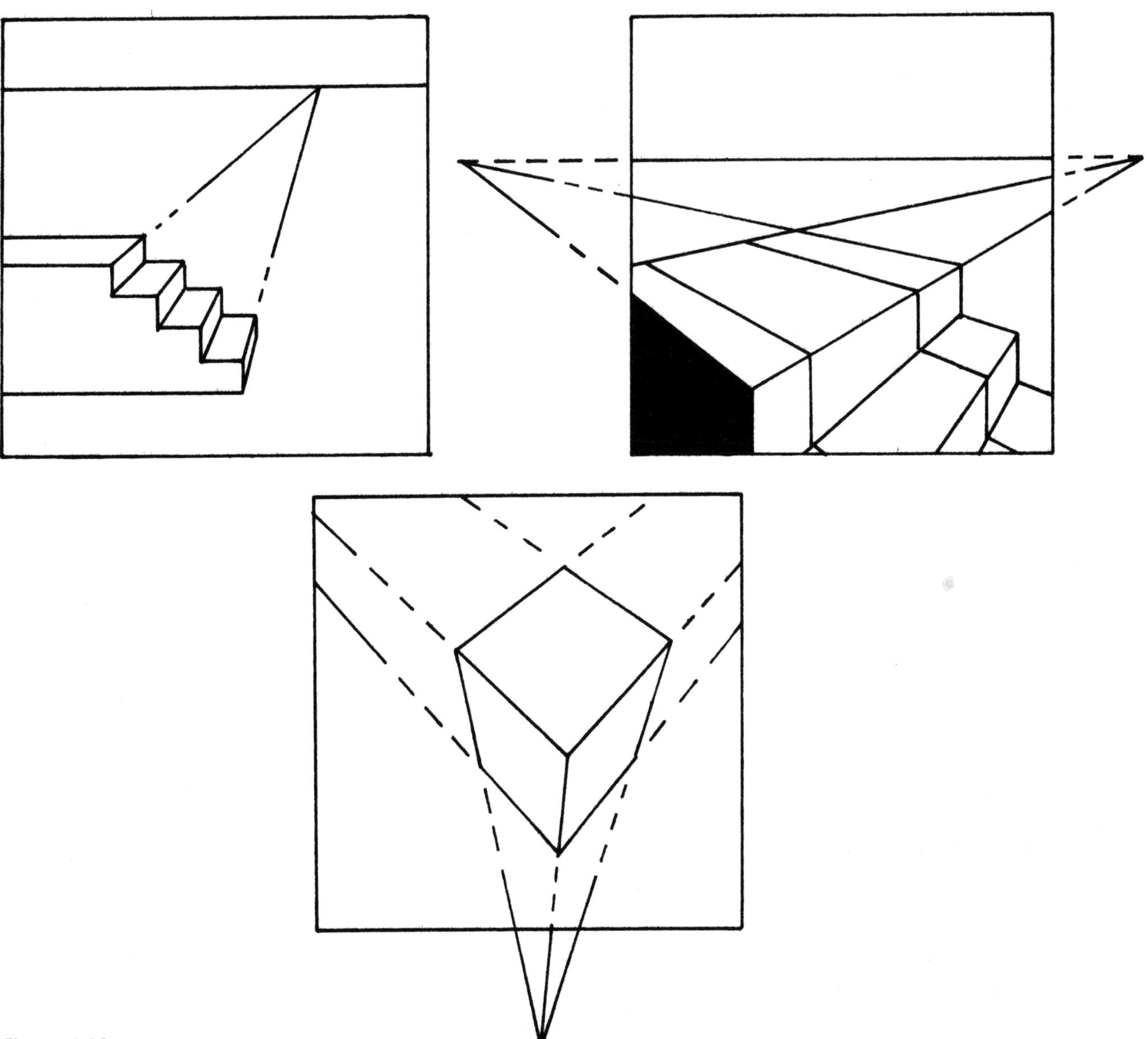

Figure 3.1B

standing in front of the scene. One set of horizontals will appear to vanish in the 'east' and the other in the 'north'. Clearly, from where you are standing, the two vanishing points should lie at right angles to each other. Now suppose that you are standing in front of a picture: imagine pointing at two points on its horizon line which are 90° apart from your point of view. This should give you some idea of how widely separated the two vanishing points in question should be placed. The rule need not be followed precisely; otherwise, pictures would only ever look correct when viewed from one spot but, if you stray too far from the principle, you will find that the drawings look distorted. As a general rule of thumb, you are unlikely to be able to squeeze both vanishing points into the picture frame; so your construction will need to extend beyond the picture frame, as shown above.

So long as you choose to view your scene by looking directly ahead, or (what amounts to the same thing) to set the picture plane vertical, all the vertical lines will appear parallel. If, however, you choose to look even slightly upwards or downwards, you will need to construct a *three-point perspective drawing* (Figure 3.1B), giving yourself three vanishing points, as shown.

Again, the problem arises as to where to locate these vanishing points in relation to each other. As in two-point perspective drawing, the main thing is to avoid putting them too close to each other. You will need to have at least two of the vanishing points outside the picture frame if you are to avoid a distorted effect. As a rough guide, remember that, when you are standing in front of the picture, the three vanishing points should represent three mutually perpendicular directions from your point of view. If you imagine tracing three such rays, you will soon see that they must be wide in relation to the size of the picture but, as mentioned before, it is not necessary to make them exactly 90° apart.

ZEEMAN'S PARADOX: WHY DOES A PICTURE LOOK RIGHT WHEN IT SHOULD LOOK WRONG?

It is worth stressing that the human eye makes sense of a wide variety of perspective pictures. For, as Professor Zeeman has pointed out, there is only one point in front of a picutre where its three mutually perpendicular vanishing points appear in mutually perpendicular directions (Figure 3.1C), and yet we happily view pictures at various distances and angles!

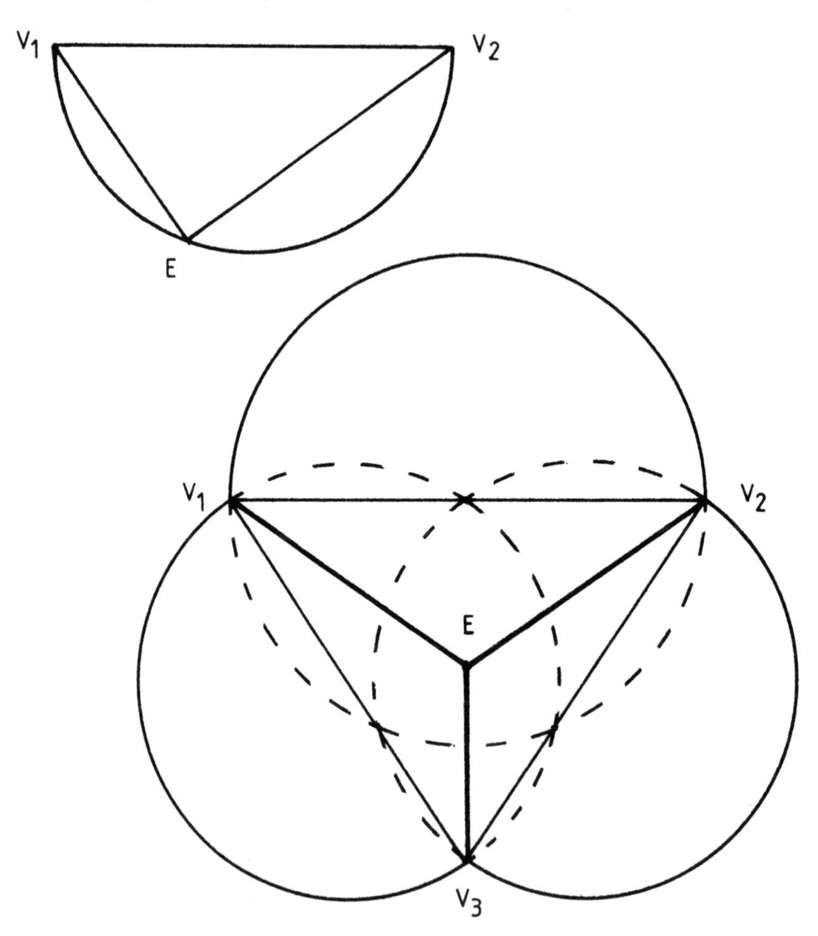

1 The angle in a semicircle is a right angle.

2 The set of all points E making a right angle with two fixed points V_1 and V_2 is the sphere on diameter V_1V_2.

3 If V_1 and V_2 are the vanishing points for two mutually perpendicular directions in a picture, and E is the viewer's eye (the viewpoint), then only when E is on this sphere will V_1 and V_2 appear to lie in perpendicular directions.

4 Three vanishing points V_1, V_2 and V_3 representing three mutually perpendicular directions will define three spheres on diameters V_1V_2, V_1V_3 and V_2V_3. Only at points common to all three spheres will V_1, V_2 and V_3 appear in three mutually perpendicular directions. There is, at most, only one such point in front of the picture plane.

Figure 3.1C

LEONARDO'S PARADOX: WHY DOES A PICTURE LOOK WRONG WHEN IT SHOULD LOOK RIGHT?

Leonardo da Vinci noted another curious thing about perspective projection (Figure 3.1D). Suppose that you are trying to depict a row of columns parallel to the picture plane and all of the same diameter. It is clear from the principle of point projection that you should depict the outer columns as wider, even though they are further away from you!

Another version of this observation is that the projections of spheres not directly ahead of you should be ellipses! At such moments the artist is obliged to overrule the mathematics to avoid apparent distortions, either by cheating or by making sure that the angle of view is quite narrow.

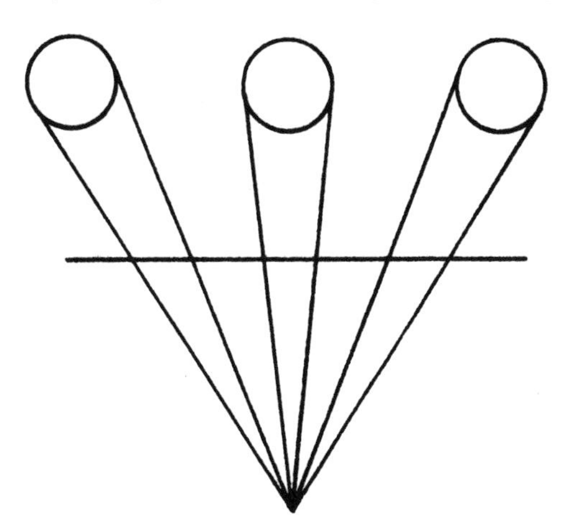

Figure 3.1D

So, each time you set about constructing a perspective drawing, you make a choice about how widely to space the mutually perpendicular vanishing points with respect to the size of the picture, which has the same effect as choosing a lens (from a wide-angle to telephoto type) in photography. An upper limit is set on the angle of view by the effects of distortion. The zero-angle or infinite-view perspective (i.e. when all vanishing points are banished to infinity, all parallels appear parallel and there is no diminution of size with recession) is an extreme case the other way. Point projection then becomes parallel projection, a system widely used by architects and engineers for simplicity and also by some oriental artists to good effect.

THE SIX-BAR GATE STORY

Our next key problem might arise as follows. Suppose that you have drawn the outline perspective view of a box-like house and wish to put windows along one of the walls, say in equally spaced intervals; is there a rule for finding the correct location for these intervals in the drawing? At its simplest, can you find the mid-point of the wall? More generally, can you find the point which divides the wall for any fraction or proportion?

Here is another example. Suppose that you have a drawing of some receding railway tracks, converging on the chosen vanishing point, and suppose that you are given the location in the drawing of any two of the regularly spaced sleepers, can you locate all the other sleepers?

The answer to both of these questions is yes. We use the *method of diagonals* for solving these problems of extending or subdividing a given receding length by a stated ratio. The simplest case is that of bisection. The diagonals of a rectangle meet in its centre (Figure 3.1E). Projections through the centre from the appropriate vanishing points allow us to bisect the rectangle length and breadth. Further bisections such as this can be performed indefinitely.

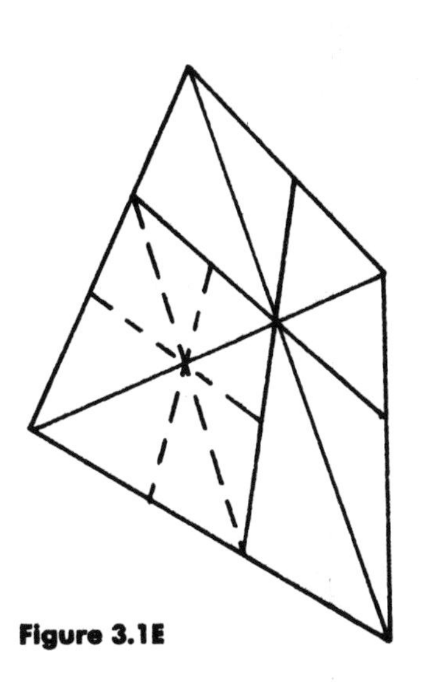

Figure 3.1E

To see how the method extends to other fractions as well, indeed to any given ratio, consider a six-bar gate drawn in perspective, as shown in Figure 3.1F. If the six bars are equally spaced, where does the diagonal

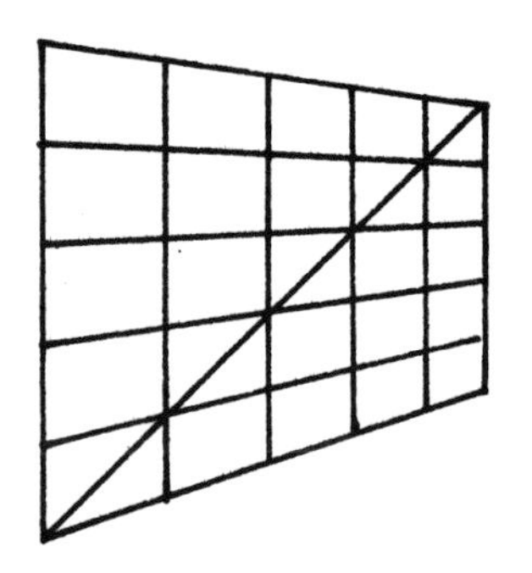

Figure 3.1F

cross-member cut each bar? The answer is at zero, one-fifth, two-fifths, three-fifths, four-fifths and five-fifths of the way across the gate. So, by constructing the verticals through the points where the diagonal cuts the bars, you can divide the receding width of the gate into five equal parts, in perspective.

What is good for fifths is good for any fraction. Figure 3.1F also shows that any receding line can be divided (externally as well as internally) by any given ratio.

ON DRAWING A SQUARE

You will soon discover that it is quite easy to construct a perspective view of a rectangular box in one-, two- and three-point perspective but that it is not always so easy to get the shape of box which you intend. For example,

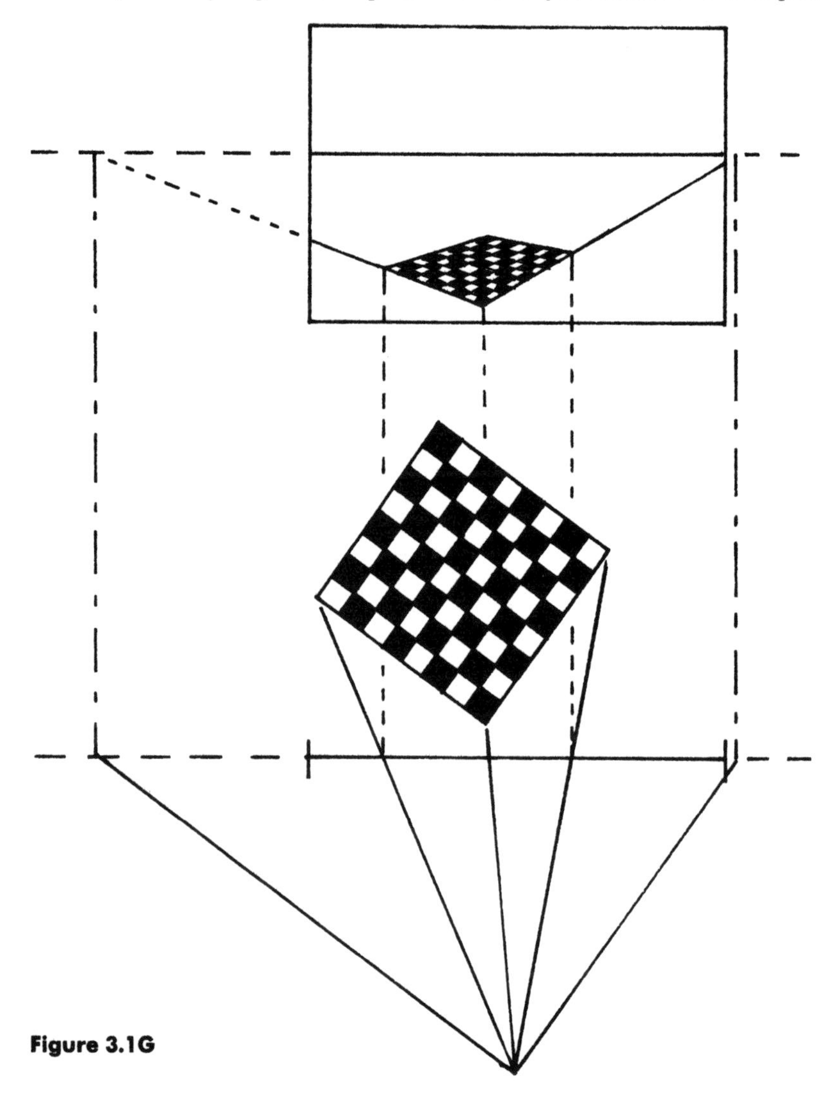

Figure 3.1G

the attempt to represent a cube is quite likely to result in a box which appears to have unequal dimensions, rectangular faces instead of square. Is there a rule for getting these proportions right in perspective?

Yes, there is, and Figure 3.1G shows the method by which a plan view of a square and projection on the picture plane is turned into a perspective view.

LOCATING FURTHER VANISHING POINTS

So far we have learnt to locate three mutually perpendicular vanishing points in the picture plane. Having located three such vanishing points, can we find a method for determining the vanishing points of lines in any other direction?

The answer is yes. We begin by looking at the simple case of diagonals. Suppose that we are given a rectangle drawn in perspective. This will establish two vanishing points of the principal perpendicular directions, as shown in Figure 3.1H, and the horizon of the rectangles plane is the line through the two vanishing points. By drawing the diagonals of the rectangle to meet the horizon we find the vanishing points of their directions. If the rectangle happens to be a square, then the diagonals point at 45° to the principal perpendiculars.

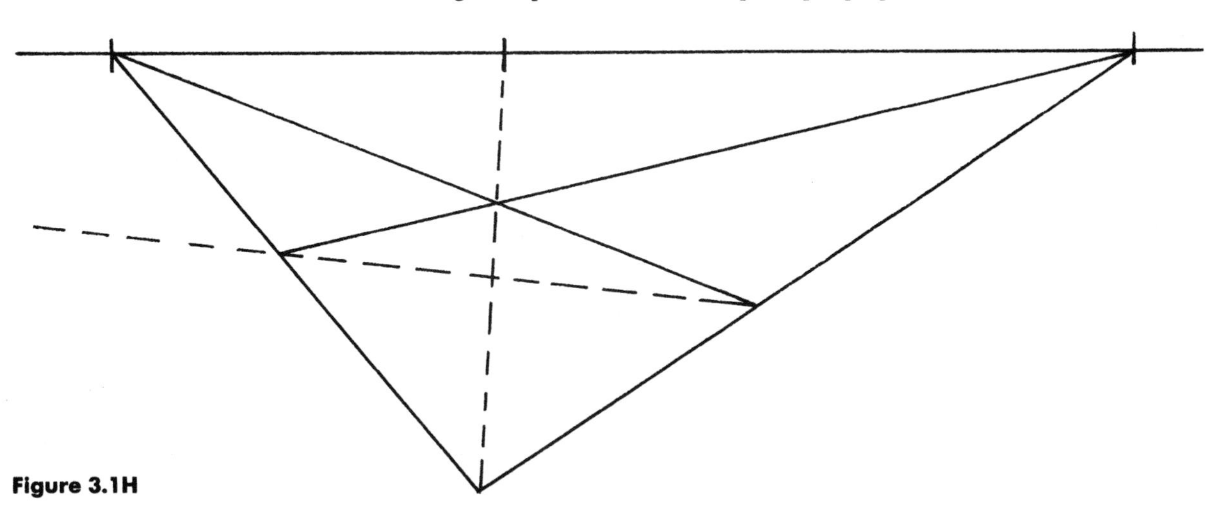

Figure 3.1H

To find any other direction in the plane of the given rectangle drawn in perspective we can use the construction in Figure 3.1I.

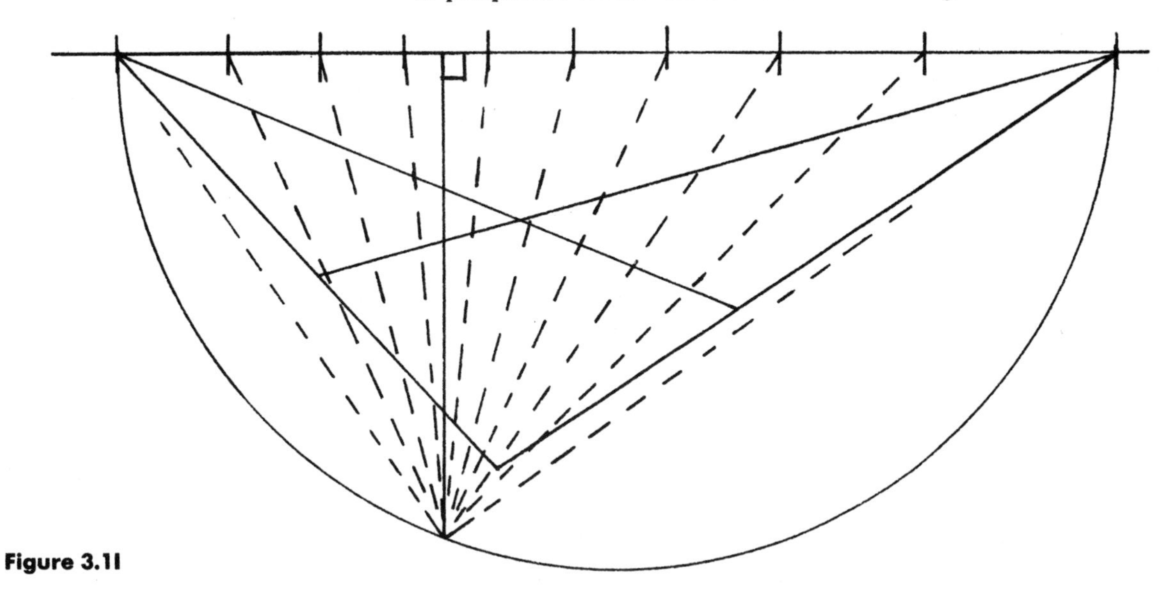

Figure 3.1I

Let the horizon line also represent the picture plane seen in plan view. To find the viewer's location in the plan, construct a semicircle on V_1V_2 as diameter. Any location on this semicircle will give a view of V_1 and V_2 in perpendicular directions, and the perpendicular bisector of the picture's width intersects this semicircle at the viewpoint. Lines through this viewpoint in any required direction intersect the horizon line at their respective vanishing points, as shown in Figure 3.1I.

CURVES IN PERSPECTIVE

The rules of perspective that we have been learning about in this chapter are all about points, lines and planes. In order to draw various curves in perspective, we must be content with approximate methods which involve a degree of judgement by eye and accuracy of handiwork. The basic method, which artists call 'gridding up' is the same for all plane curves and is illustrated in Figure 3.1J for a circle.

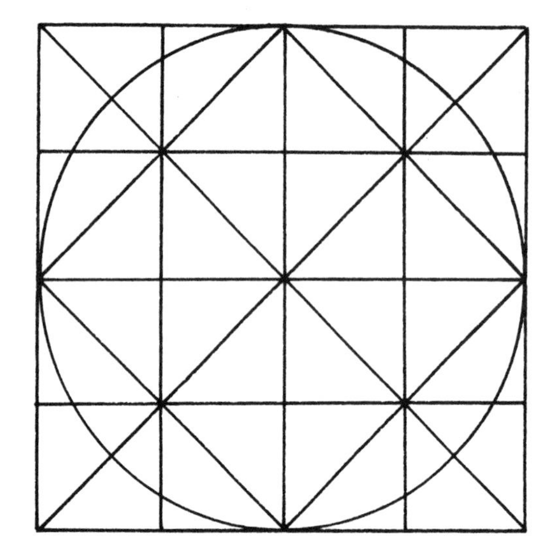

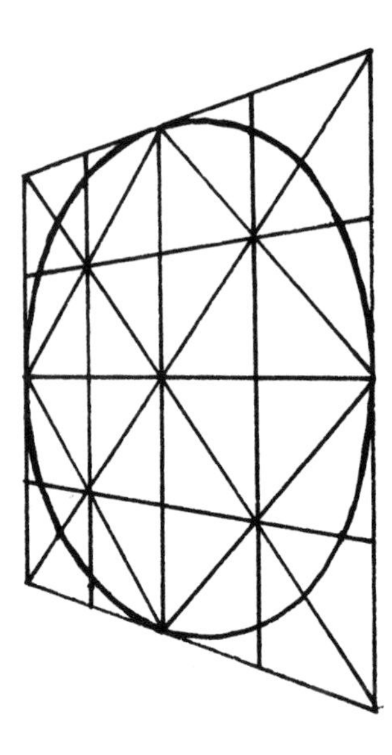

Figure 3.1J

Method

Start by drawing the curve not in perspective, together with a superimposed square grid, and shown in Figure 3.1J. Next, draw the square grid in perspective. Then, point by point, see where the curve not in perspective cuts its grid and mark the corresponding intersections on the perspective grid. Finally, join these intersections to draw the curve in perspective.

SHADOWS AND REFLECTIONS

Various problems involving shadows cast by sunlight or lamplight, and mirror reflections in glass or still water, present many interesting exercises, which will be left for you to explore and solve. Clues for dealing with shadows: locate the point or direction of the light source and project lines from there through points of the object, casting the shadow to the surface on which shadow falls. Clues for dealing with reflections: establish where the plane of reflection (mirror or water surface) cuts the object to be reflected, and then construct a 'Siamese twin' of the object, joined in this plane and drawn in the same perspective.

Exercise 6

Explore all these problems and any other perspective drawing with preliminary sketches by hand, before using a ruler for the actual construction.

1 Practise drawing by quick freehand sketches various rectangles, squares and boxes in one-point, two-point and three-point perspective. Experiment with various locations and relations for the vanishing points with respect to the picture frame.

2 Construct a square in two-point perspective, and then divide it into a grid of 8 × 8 equal squares, as on a chessboard.

3 Construct a box-like house with a flat roof, and also a box-like house with a pitched roof, with windows and doors in both cases, in two-point perspective.

4 Repeat question **3** using three-point perspective.

5 Construct a railway track and signal box in two-point perspective, showing at least ten of the equally spaced sleepers.

6 Construct a two-point perspective of a straight canal with a bridge on a still day.

7 Construct a box on a plane surface together with its shadow cast by the Sun. Construct a similar box with its shadow cast by a nearby lamp. (Note that the light is treated as a point source for simplicity in both cases.)

8 Construct one or more circles in perspective.

9 Construct a window in two-point perspective with a semicircular arched top.

10 Construct a rectangular staircase in two-point perspective.

11 Construct a spiral staircase in two-point perspective. (This problem should only be attempted when all the preceding exercises have been fully mastered.)

12 Repeat questions **10** and **11** in three-point perspective.

13 Construct an infinite-view perspective of any of the drawings for questions **1–12**.

14 Look in art books and galleries to find pictures by any of the following artists to see how they employed the rules of perspective:

Jan Van Eyck (1390–1440)
Paolo Uccello (1397–1475)
Masaccio (1401–1428)
Piero della Francesca (1416–1492)
Andrea Mantegna (1431–1506)
Leonardo da Vinci (1452–1519)
Albrecht Dürer (1471–1528)

Can you find examples of one-point, two-point and three-point perspectives?

15 Find other examples of art from the fifteenth century and earlier times which show a less successful understanding of perspective.

16 Look for examples in the graphic works of M.C. Escher which show his mastery of perspective. Find good examples of three-point perspective.

17 Repeat question **16**, but this time look for the use of novel kinds of perspective, such as cylindrical perspective, where the scene is projected onto a cylinder surrounding the viewer and then unrolled flat.

18 Use a six-bar gate diagram, or something comparable, and then, by using arguments from similar triangles, show that an arithmetic sequence of lengths OA, OB, OC, ... in the scene becomes a harmonic sequence of lengths in the perspective drawing. *Note*: an *arithmetic* sequence of numbers $n_1, n_2, n_3, \ldots$ is one in which $n_1 - n_2 = n_2 - n_3 = n_3 - n_4$ etc.; a *harmonic* sequence of numbers is given by $c/m_1, c/m_2, c/m_3, \ldots$ where $m_1, m_2, m_3, \ldots$ form an arithmetic sequence.

Reading

Escher, 1972; Gombrich, 1964; Kline, 1967, 1972; Parramon, 1984; Pedoe, 1976; de Vries, 1968.

4 The Story of Trigonometry

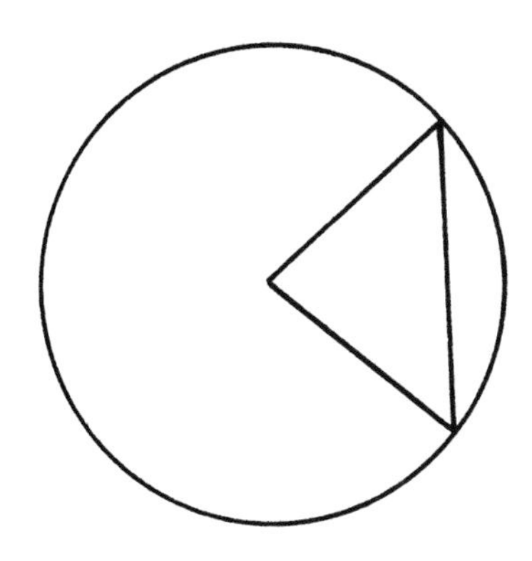

Figure 4.0A

In one way or another, most of this book is about the circle and the straight line. What happens when they come together? As with two partners in a dance or two leading characters in a murder mystery, we have to work out how they relate to each other (Figure 4.0A). How does the measure of length relate to the measure of angle? In a circle of known diameter, what is the length of the circumference? What is the length of a chord which subtends a given angle at the centre of the circle?

These problems were solved in a systematic manner by the development of Greek trigonometry, which includes the theorem of Pythagoras, the use of polygons (Drawing **34**), the use of sine and cosine ratios in right-angled triangles (Drawing **35**) and the use of $\pi = 3\frac{1}{7}$ (or its sharper modern value of 3.141 59) (Drawing **36**) in such formulae as $C = \pi D$ and $A = \pi R^2$. All this, of course, is now basic to school mathematics.

It is quite easy to learn how to use each of these parts of trigonometry without needing to understand how they are connected together or how each was discovered. Today we obtain the square root of any given number, or the sine of any given angle, at the press of a calculator button; before the invention of electronic computers, we would look them up in ready-made tables of values. While this achievement is a great convenience to users of mathematics, it does hide from view the path of discovery and string of calculations which leads from Pythagoras' theorem to Archimedes' computation of π and to the computation of the sine and cosine tables by Hipparchus and Ptolemy. Yet this story is well worth telling, at least in mathematical outline. It provides a contrasting companion to the story of Euclid's geometry in which there is no mention of π or of sines. The story of trigonometry is about the power of Pythagoras' theorem used in repeated calculations to obtain systematically increasingly better approximations for lengths, especially in problems involving circles and angles.

Reading

Aaboe, 1964; Abbott, 1940; Boyer, 1968; Courant and Robbins, 1941; Hogben, 1967; Resnikoff and Wells, 1984.

34 Hexagon, 12-gon, 24-gon, 48-gon, 96-gon

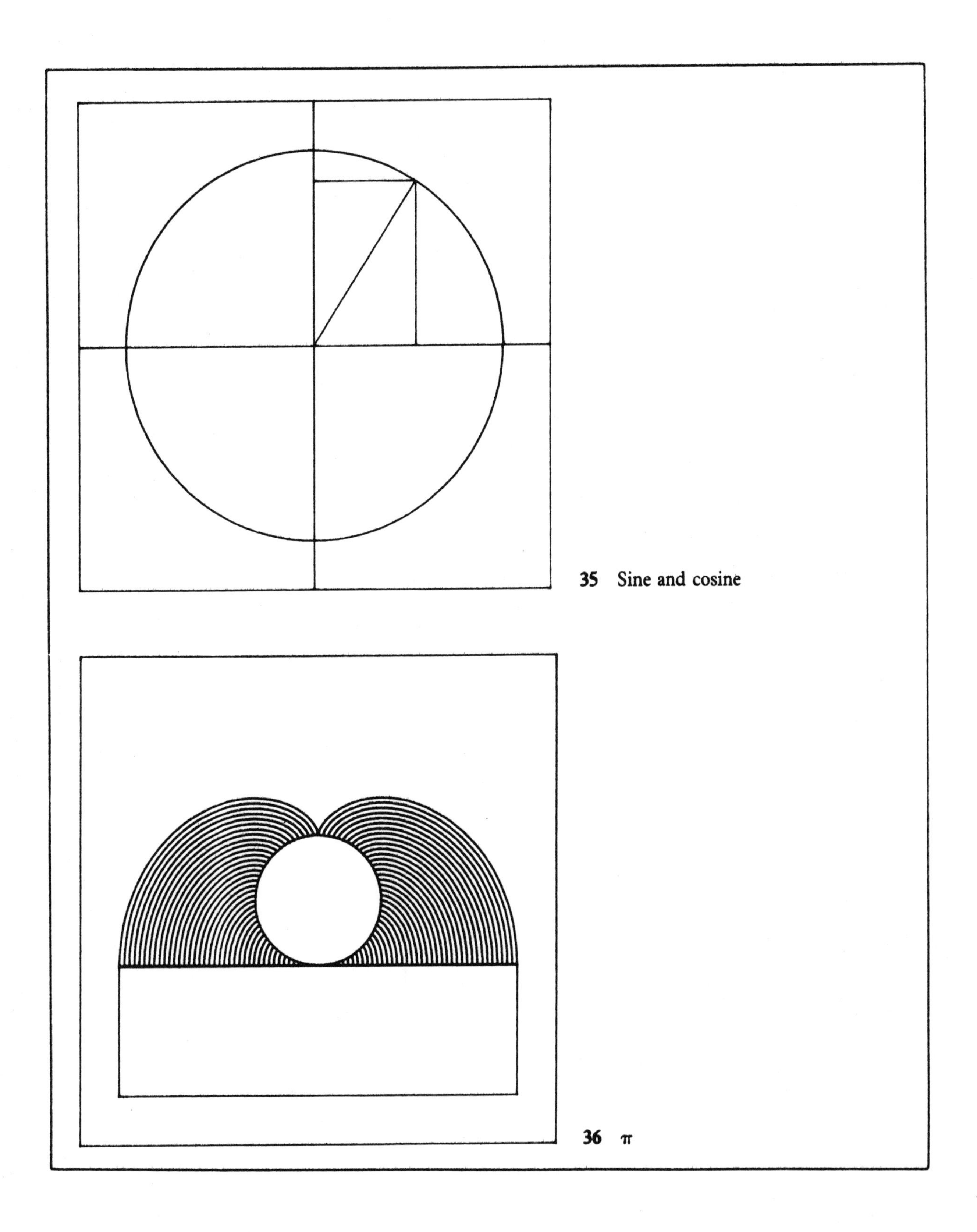

35 Sine and cosine

36 π

4.1 The theorem of Pythagoras

How long is the diagonal of a rectangle whose side lengths are a and b respectively? This is solved by using Pythagoras' theorem, because the diagonal is the *hypotenuse* of a right-angled triangle (Figure 4.1A).

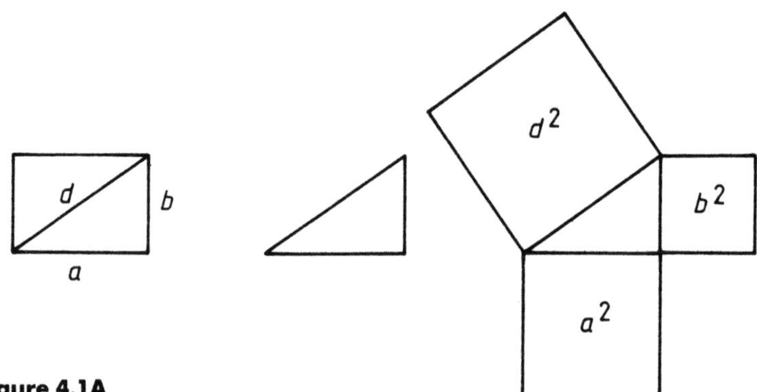

Figure 4.1A

Pythagoras' theorem

$$d^2 = a^2 + b^2$$

Therefore,

$$d = \sqrt{(a^2 + b^2)}$$

As the coordinate axes of the Cartesian x–y system are perpendicular to each other, the theorem of Pythagoras enables us to compute the distance between any two points $P(x_1, y_1)$ and $Q(x_2, y_2)$ in the plane (Figure 4.1B).

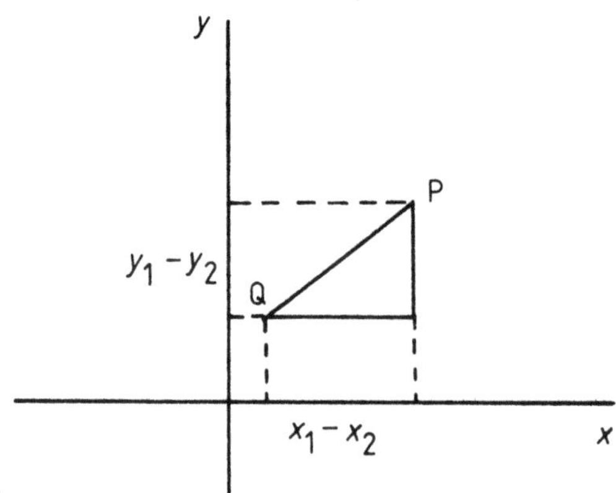

Figure 4.1B

$$d = \sqrt{\{(x_1-x_2)^2 + (y_1-y_2)^2\}}$$

This rule will be very useful in computer graphics, where points are located by Cartesian coordinates.

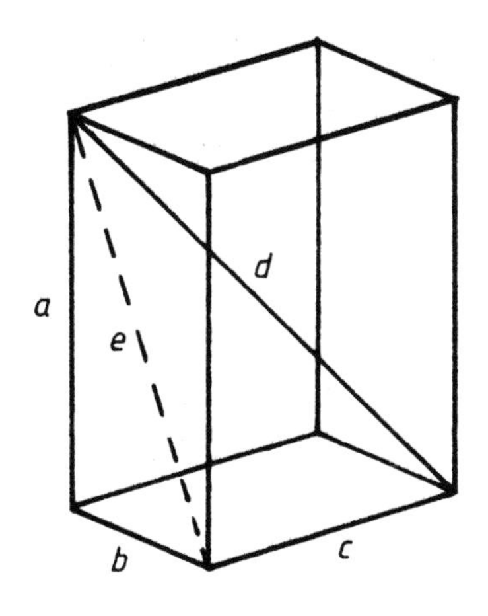

Figure 4.1C

What is the length of a diagonal in a rectangular box whose sides measure a, b and c respectively? Again, Pythagoras' theorem gives us the answer:

$$d = \sqrt{(a^2 + b^2 + c^2)}$$

This can be proved using Figure 4.1C:

$$\begin{aligned} e^2 &= a^2 + b^2 \\ d^2 &= e^2 + c^2 \\ &= a^2 + b^2 + c^2 \end{aligned}$$

thus providing the formula for finding the distance between any two points in three-dimensional space $P(x_1, y_1, z_1)$ and $Q(x_2, y_2, z_2)$:

$$d = \sqrt{\{(x_1-x_2)^2 + (y_1-y_2)^2 + (z_1-z_2)^2\}}$$

Exercise 7

1 Find the distance PQ in each of the following examples.
- **a** P(3, 7) and Q(4, 2).
- **b** P(1, 2, 3) and Q(3, 4, 5).
- **c** P(0, 2, 3) and Q(3, −2, 0).

2 Find the diagonal lengths of the following rectangles.
- **a** $a = 1$ and $b = 2$.
- **b** $a = 1$ and $b = \sqrt{2}$.
- **c** $a = b = 1$.

3 Find the height of an equilateral triangle of side length equal to 1.

4 Find the heights of the following isosceles triangles.
- **a** Base of 2 and two equal sides of 3.
- **b** Base of 1 and two equal sides of 2.
- **c** Base of 2 and two equal sides of $\sqrt{3}$.

5 Find the length of the following figures if each of these figures is circumscribed by a circle of radius 1.
- **a** The side of a square.
- **b** The diagonal AC of a regular hexagon ABCDEF.

6 Find the length of the diagonal of a square box of side length equal to 1.

7 Find the height of a square pyramid whose sides are all equal to 1.

8 Find the ratios of the three sides of the triangle formed by the three centres of curvature in the following (see Section 1.1).
- **a** Moss' egg.
- **b** Thom's egg.
- **c** Cundy and Rollett's egg.

HOW OLD IS THE THEOREM?

According to legend, Pythagoras was a Greek prophet and mystic of the sixth century BC who travelled to Egypt and to Babylon, and perhaps also to India, where he acquired his knowledge of mathematics and science, and who returned to form a school of learning which was to inspire all of Greek Science, and hence all modern science. We know from clear evidence of its use in these earlier cultures that Pythagoras did not discover the theorem, but he is supposed to have discovered the first *proof*. Such a feat of logical argument was to become the distinctively new achievement of Greek mathematics.

PYTHAGOREAN TRIPLETS

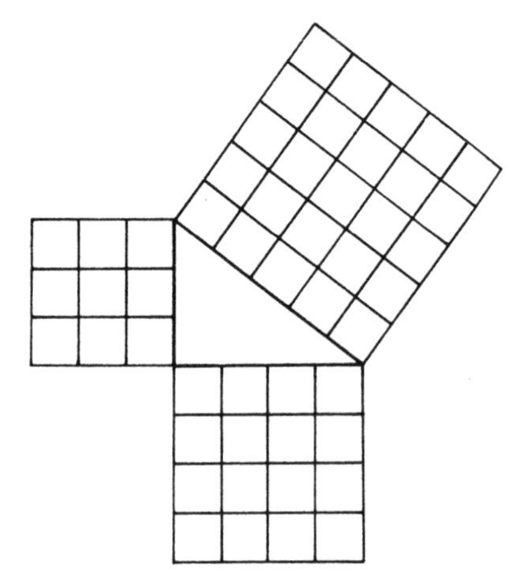

Figure 4.1D

The earliest knowledge of the theorem probably came from the observation that a triangle whose sides are 3, 4 and 5 units in length forms a right-angled triangle, together with the observation that $3^2 + 4^2 = 5^2$ (Figure 4.1D).

Any three whole numbers a, b and c which satisfy the rule $a^2 + b^2 = c^2$ are called *Pythagorean triplets*. It so happens that there are infinitely many such triplets. Here are the first few:

3, 4, 5	
5, 12, 13	8, 15, 17
7, 24, 25	12, 35, 37
9, 40, 41	16, 63, 65
...	...

A triangle formed by lengths of any of these triplets will be right angled. Babylonian tablets containing lists of Pythagorean triplets have been discovered which date back to 1900–1600 BC.

Exercise 8

1 Find the next two Pythagorean triplets in the following.
  a The left-hand column above.
  b The right-hand column above.

PROOFS OF PYTHAGORAS

Knowing how to use the rule of Pythagoras is quite a different matter from, and quite a lot easier than, knowing how to *prove* it. A proof must be logical, general and as elegant as possible. We do not know what Pythagoras' proof was. The earliest to come down to us is that given by Euclid. Figure 4.1E shows four of the most elegant proofs, including Euclid's. In each case the diagram is provided but the details of the argument are left to you to find, using the following clues.

The Chinese proof is probably the clearest, if you know that two right-angled triangles make a rectangle and that $(a + b)^2 = a^2 + 2ab + b^2$.

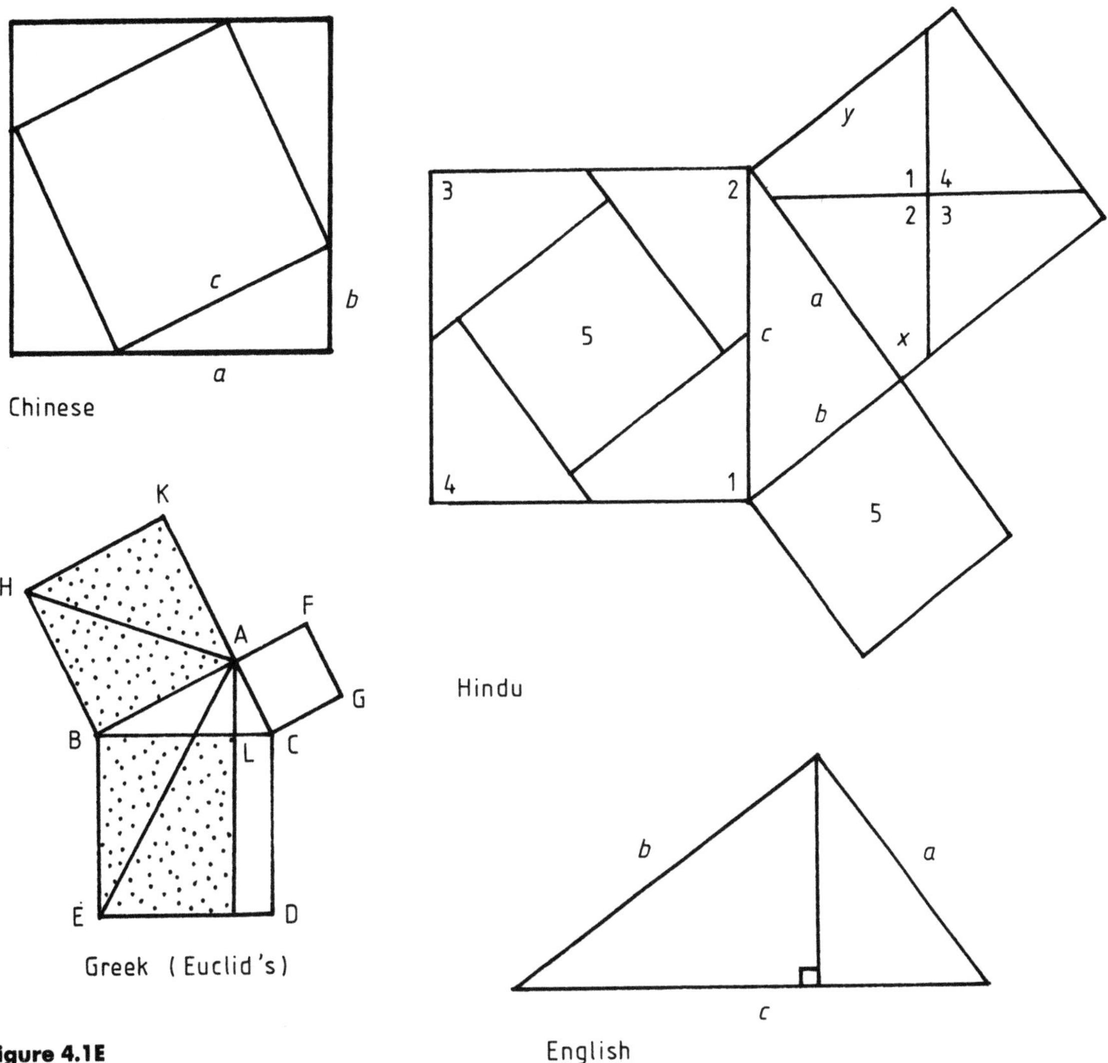

Figure 4.1E

The English proof is neat but requires you to argue with similar triangles.

The Hindu proof is an argument by dissection and requires you to find the rule for cutting the middle-sized square.

Euclid's proof compares the area of each of the smaller squares with a part of the largest square via congruent triangles.

4.2 Square roots

It will be clear by now that the use of Pythagoras' theorem entails the problem of finding square roots. (A square root is the number which, when multiplied by itself, makes a square, e.g. 7 is the square root of 49.) Some whole numbers are perfect squares: 1, 4, 9, 16, 25, 36, ... The square roots of these numbers are easy to find: 1, 2, 3, 4, 5, 6, ... However, for all other numbers the square root is not a whole number, nor is it a fraction. The best that we can do is to obtain a sufficiently good approximation for practical purposes. Systematic methods for doing this were used by the ancient Babylonians. The method which follows is essentially theirs.

Problem To find $\sqrt{x}$, where x is a whole number, but not a perfect square, e.g. to find $\sqrt{10}$.

Method **a** Find the nearest perfect square, in this case 9, and let $\sqrt{10} = 3 + d$.

b Therefore,

$$10 = (3 + d)^2 = 9 + 6d + d^2$$

so that

$$1 = d(6 + d)$$

and

$$d = \frac{1}{6 + d}$$

c Use this equation as an iteration formula to obtain increasingly better approximations for d. Start with an initial guessed estimate for d called d_0 and proceed as follows:

$$d_1 = \frac{1}{6 + d_0}, \; d_2 = \frac{1}{6 + d_1}, \text{ and so on}$$

In our example of finding $\sqrt{10}$, if we start with $d_0 = 1$, this gives $d_1 = 1/7$, which gives $d_2 = 7/43$, which gives $d_3 = 43/265$, and so on. The sequence of approximations converges quite rapidly to the true value of d, and we see that

$$\sqrt{10} \approx 3\frac{43}{265}$$

a sufficiently close approximation for practical purposes.

IRRATIONAL NUMBERS

A number which cannot be expressed exactly as a ratio of whole numbers is called *irrational*. (You should think of this as meaning 'not a ratio of whole numbers' rather than 'mad'.) All square roots, apart from the roots of perfect squares, are irrational. So are $\pi = 3.141\,59\ldots$ and $e = 2.718\ldots$. The best that we can do is to offer a near enough approximation.

The proof that $\sqrt{2}$ is irrational is another of the triumphs attributed to Pythagoras. It may have run approximately as follows.

- **a** Every number can be expressed as a product of prime numbers, and the product will be unique for each number. For example $8 = 2 \times 2 \times 2$, $15 = 3 \times 5$, $24 = 2 \times 2 \times 2 \times 3$, and so on.
- **b** Perfect squares have an even number of prime factors. For example $25 = 5 \times 5$, $36 = 2 \times 3 \times 2 \times 3$, and so on.
- **c** So, if $\sqrt{2} = p/q$, then $2q^2 = p^2$.
- **d** However, this is impossible, because the left-hand side of this equation has an odd number of prime factors while the right-hand side has an even number of prime factors.

This discovery is supposed to have shaken the Pythagoreans who placed much importance on whole numbers, but we no longer worry about this and have the powerful system of decimal notation to help us.

Two lengths are called *incommensurate* if no choice of units allows both to be expressed as a whole number of units. Examples are as follows.

- **a** The side and the diagonal of a square.
- **b** The side and diagonal of a hexagon.
- **c** The diameter and circumference of a circle.

Exercise 9

1 Use the iterative method to find $\sqrt{2}$, $\sqrt{3}$, $\sqrt{5}$, $\sqrt{7}$.

2 Prove that the following are irrational.
a $\sqrt{3}$.
b $\sqrt[3]{7}$.

4.3 The story of pi (π)

How far is it round a circle? How far does a wheel travel in rolling one revolution? The answer in each case is, of course,

$$\text{circumference} = \pi \times \text{diameter}$$

For all practical purposes, we can safely use $\pi = 3.1416$, the value used by Ptolemy (circa AD 150). It is not necessary to understand how π is calculated in order to use it, only that it is the ratio of circumference to diameter in a circle. However, for those wishing to know a little more about this most famous and elusive number the story of its calculation is well worth telling.

The earliest estimates for the value of π probably began with the crude approximation of $\pi = 3$. The Egyptians used the value of $4(8/9)^2 = 3.1605$, while the Babylonians used a slightly closer value of $25/8 = 3.125$. However, it was Archimedes (287–212 BC) who effectively solved the problem of computing the value of π. By calculating the perimeters of two regular 96-gons, one inscribed and the other exscribed on a circle of unit diameter, he was able to conclude that

$$3\frac{1}{7} > \pi > 3\frac{10}{71}$$

Starting with a regular hexagon, he constructed first a regular 12-gon by bisecting the six arcs of the circle, then a regular 24-gon in the same manner, then a regular 48-gon and, finally, the regular 96-gon. At each stage, the perimeter of the regular polygon is obtained from the perimeter of the previous regular polygon by means of Pythagoras' theorem. If a greater accuracy for π is required, simply continue the process of bisection and calculation. At each stage the true circle lies between the inscribed polygon, which is shorter, and the exscribed polygon, which is longer. The greater the number of sides, the closer it is to the length of the true circle.

To see how the computation works out, consider a circle of unit diameter and any arc of this circle sitting on a chord of length CH1. We then bisect this arc and compute the length CH2 of the chord of one-half of this arc. The computation involves using Pythagoras' theorem twice (Figure 4.3A).

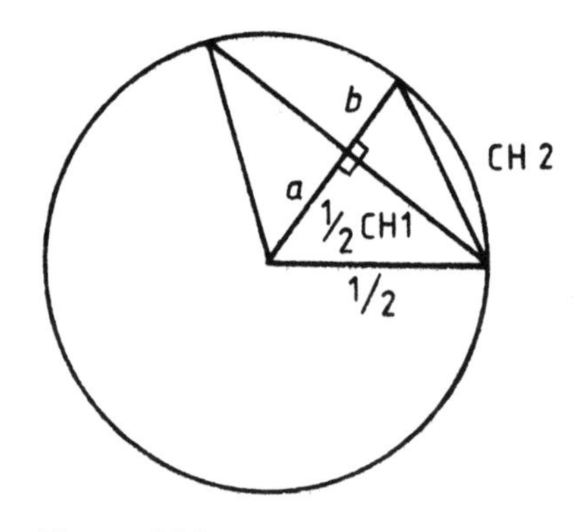

Figure 4.3A

$$a^2 + (\tfrac{1}{2}\text{CH1})^2 = (\tfrac{1}{2})^2,$$

giving

$$a = \tfrac{1}{2}\sqrt{\{1 - (\text{CH1})^2\}}$$

Next

$$\begin{aligned} b &= \tfrac{1}{2} - a \\ &= \tfrac{1}{2} - \tfrac{1}{2}\sqrt{\{1 - (\text{CH1})^2\}} \end{aligned}$$

and

$$\begin{aligned}(CH2)^2 &= b^2 + (\tfrac{1}{2}CH1)^2 \\ &= \tfrac{1}{4} - \tfrac{1}{2}\sqrt{\{1 - (CH1)^2\}} + \tfrac{1}{4}\{1 - (CH1)^2\} \\ &\quad + (\tfrac{1}{2}CH1)^2 \\ &= \tfrac{1}{2} - \tfrac{1}{2}\sqrt{\{1 - (CH1)^2\}}\end{aligned}$$

This process of bisection can be repeated indefinitely to obtain

$$\begin{aligned}(CH3)^2 &= \tfrac{1}{2} - \tfrac{1}{2}\sqrt{\{1 - (CH2)^2\}} \\ &= \tfrac{1}{2} - \tfrac{1}{2}\sqrt{[\tfrac{1}{2} + \tfrac{1}{2}\sqrt{\{1 - (CH1)^2\}}]} \text{ and so on.}\end{aligned}$$

Putting CH1 = $\frac{1}{2}$, the side length of an inscribed regular hexagon, then CH5 gives the side length of an inscribed regular 96-gon, whose perimeter is therefore

$$96\sqrt{(\tfrac{1}{2} - \tfrac{1}{2}\sqrt{[\tfrac{1}{2} + \tfrac{1}{2}\sqrt{\{\tfrac{1}{2} + \tfrac{1}{2}\sqrt{(\tfrac{1}{2} + \tfrac{1}{2}\sqrt{\tfrac{3}{4}})}\}}]})}$$

Exercise 10

1 Find the perimeter lengths to 4 decimal places of the following.
a Inscribed 96-gon.
b Circumscribed 96-gon.

AREA OF A CIRCLE

Archimedes' method of approximating a circle to a many-sided regular polygon also provided the formula for the area of a circle:

$$\text{area} = \pi r^2$$

For example, the 96-gon can be sliced into 96 equilateral triangles whose height approaches the radius of the circle r, and whose total base length approches $2\pi r$.

RADIAN MEASURE

The Babylonians, whose pre-decimal arithmetic relied on bases of 6 and 60, divided the circle into 360°, a system of angular measure which is still the most commonly used and understood of several alternatives. It is important to realise that there is nothing intrinsic to the number 360 which relates it to the circle, but only the arithmetical convenience of working in whole numbers and allowing whole number division by 2, 3, 4, 5, 6, 8, 9, 10, 12, 15, 18, 20, 24, 30, 36, 40, 45, 48, 60, 90, 120 and 180.

In engineering we count revolutions, giving the unit of circular measure as 360° = 1 turn. If we used this system, 90° would be called $\frac{1}{4}$, and the sum of the angles in a triangle would be $\frac{1}{2}$, and so on. Pocket calculators usually have a button which allows you to switch between three alternatives for angular measure: degrees, gradians and radians, where

$$360° = 400 \text{ gradians} = 2\pi \text{ radians}$$

Radian measure is particularly useful in higher mathematical analysis and is frequently the 'natural' system of angular measure employed in computing. Thus, if you command your computer to calculate SIN(60), thinking that you are about to get the sine of 60°, it is most likely that it will calculate the sine of 60 radians, which is a very different angle. For the purposes of conversion from one system to the other the following formulae should be used:

$$X \text{ degrees} \quad = 2\pi X/360 \text{ radians}$$

and

$$Y \text{ radians} \quad = 360Y/2\pi \text{ degrees}$$

The principal advantage of radian measure lies in the fact that the length of an arc in a circle of unit radius which subtends an angle of A radians at the centre is simply A (Figure 4.3B). For a circle of radius R,

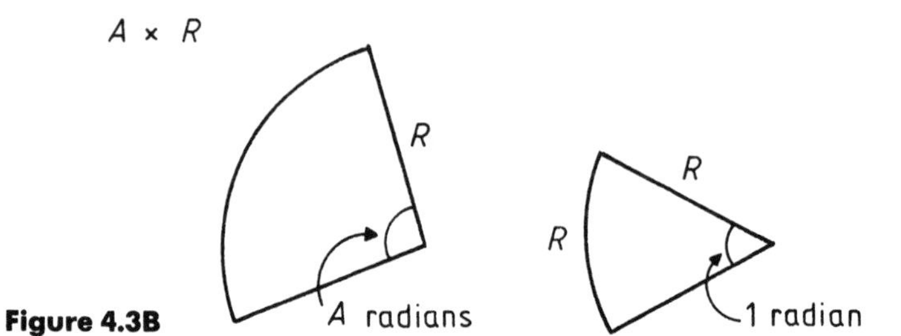

Figure 4.3B

$$\text{arc length} = A \times R$$

One radian is the part of a circle for which the arc length is equal to the radius and is an angle of approximately 57.3°.

FURTHER APPROACHES TO THE VALUE OF π

The following formulae are strictly for the collector, but they do give an idea of how elusive this number is. Not only is π *irrational* (i.e. like $\sqrt{2}$ it cannot be given as a ratio of finite whole numbers), but also it is *transcendental* (which means that not even powers of π or combinations of powers of π give finite whole numbers). So all formulae for computing π are themselves infinitely long.

Archimedes' solution can be expressed as follows:

$$\pi = \lim_{m \to \infty} [2^m \sqrt{}(2 - \sqrt{}[2 + \sqrt{}\{2 + \sqrt{}(2 + \ldots + \sqrt{2})\}])]$$

with m nested square roots.

A particularly elegant formula is given by the infinite sum

$$\frac{\pi^2}{6} = 1 + \frac{1}{4} + \frac{1}{9} + \frac{1}{16} + \frac{1}{25} + \ldots$$

Viete (1540–1603) found the infinite product

$$\frac{2}{\pi} = \sqrt{\frac{1}{2}} \times \sqrt{\left(\frac{1}{2} + \frac{1}{2}\sqrt{\frac{1}{2}}\right)} \times \sqrt{\left\{\frac{1}{2} + \frac{1}{2}\sqrt{\left(\frac{1}{2} + \frac{1}{2}\sqrt{\frac{1}{2}}\right)}\right\}} \times \ldots$$

Another infinite product was found by Wallis (1676–1703):

$$\frac{\pi}{2} = \left(\frac{2}{1} \times \frac{2}{3}\right) \times \left(\frac{4}{3} \times \frac{4}{5}\right) \times \left(\frac{6}{5} \times \frac{6}{7}\right) \times \left(\frac{8}{7} \times \frac{8}{9}\right) \times \left(\frac{10}{9} \times \frac{10}{11}\right) \times \left(\frac{12}{11} \times \frac{12}{13}\right) \times \dots$$

Probably the simplest formula is due to Leibnitz (1646–1716):

$$\frac{\pi}{4} = 1 - \frac{1}{3} + \frac{1}{5} - \frac{1}{7} + \frac{1}{9} - \frac{1}{11} + \dots$$

Brouncker (1620–1687) found the following continued fraction:

$$\frac{4}{\pi} = 1 + \cfrac{1}{2 + \cfrac{9}{2 + \cfrac{25}{2 + \cfrac{49}{2 + \cfrac{81}{2 + \dots}}}}}$$

Finally, there are the *principal convergents* of π: the sequence of reduced fractions with increasing denominators which successively approach closer and closer to the true value of:

$$\frac{3}{1}, \frac{22}{7}, \frac{333}{106}, \frac{355}{113}, \frac{102573}{32650}, \frac{102928}{32763}, \dots$$

Exercise 11

1 Give the next term in the following.
  a Wallis' product.
  b Leibnitz' series.

2 Using a calculator, obtain approximations to π with each of the first five formulae above.

3 Repeat question **2**, using a computer and writing a program.

4.4 The story of sine and cosine

Before the invention of electronic computers, ready-made tables of sine and cosine values were used. Normally, these tables would show a value for every angle between 0° and 90° at regular intervals of one-sixtieth of a degree. Here is a fragment of such a table, for intervals of every degree.

A	$\sin A$	$\cos A$
10°	0.1736	0.9848
11°	0.1908	0.9816
12°	0.2079	0.9781
13°	0.2250	0.9744
14°	0.2419	0.9703
15°	0.2588	0.9659

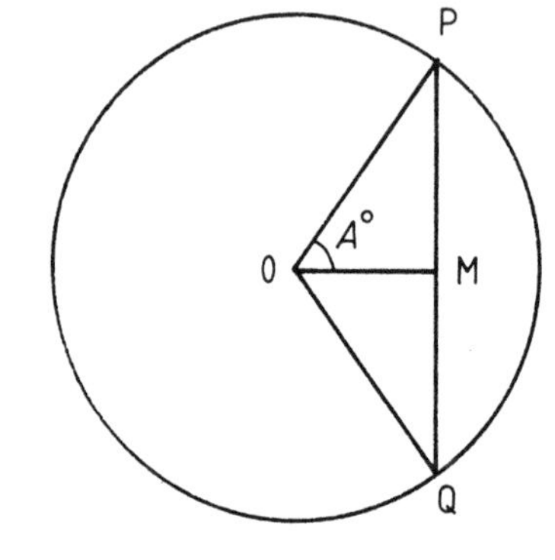

Figure 4.4A

To get a picture of what they mean, consider Figure 4.4A which shows a circle of unit radius. A chord PQ makes an isosceles triangle with its centre O. The perpendicular bisector of the chord, OM, bisects the angle subtended at the centre by the chord. Call each half-angle A°. Then,

$$\sin A = \text{PM}$$
$$\cos A = \text{OM}$$

(and chord $2A$) = PQ = $2\sin A$). As A is made to vary smoothly between 0° and 90°, it is easy to see how $\sin A$ will vary smoothly between 0 and 1, while $\cos A$ will vary smoothly between 1 and 0.

Consider next the path of a point P(x, y) circling the origin of a Cartesian coordinate system in the plane with a radius of 1. As P completes its circuit, it is easy to see how its x and y values vary smoothly between -1 and $+1$ accordingly. If A is the angle between OP and the x axis measured in an anticlockwise direction, then (Figure 4.4B)

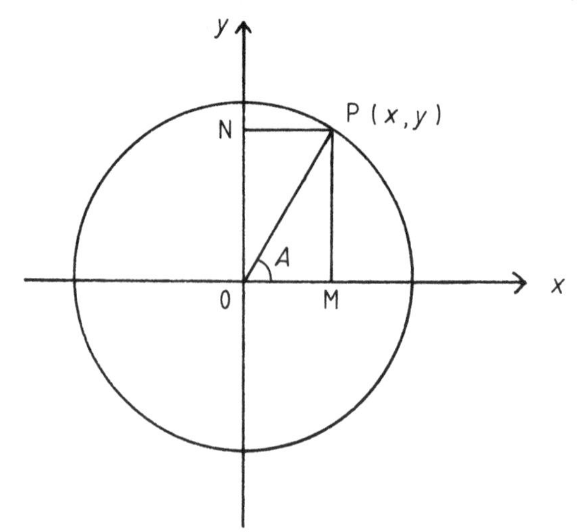

Figure 4.4B

$$\sin A = y = \text{ON}$$
$$\cos A = x = \text{OM}$$

By making a graph of sin A as a function of A as A varies, you obtain the *sine wave* (Figure 4.4C). (The cosine wave is exactly the same, except that it is displaced by 90° to the left.)

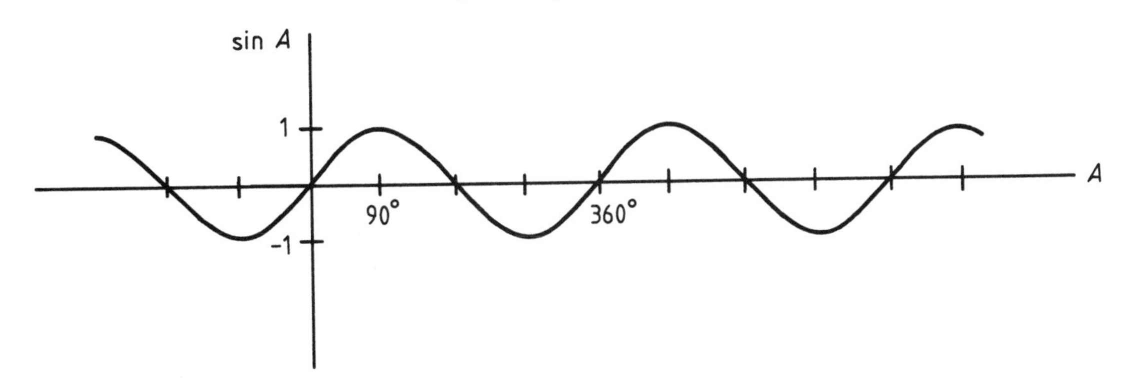

Figure 4.4C

Dispensing with the circle for a moment, and concentrating on the righ-angled triangle (Figure 4.4D), we see that

Figure 4.4D

$$\sin A = \frac{\mathrm{MP}}{\mathrm{OP}}$$

$$\cos A = \frac{\mathrm{OM}}{\mathrm{OP}}$$

and

$$\tan A = \frac{\mathrm{PM}}{\mathrm{OM}} = \frac{\sin A}{\cos A}$$

Because of the theorem of Pythagoras, and the angle sum of a triangle, the following formulae can all be easily proved:

$$\begin{aligned}
\sin^2 A + \cos^2 A &= 1 \\
\sin A &= \cos(90° - A) \\
\cos A &= \sin(90° - A) \\
\cos A &= \cos(-A) \\
\sin A &= \sin(180° - A) \\
\sin A &= -\sin(-A) \\
\sin A &= \sin(A + 360n) \text{ for } n = 0, \pm 1, \pm 2, \pm 3, \ldots \\
\cos A &= \cos(A + 360n) \text{ for } n = 0, \pm 1, \pm 2, \pm 3, \ldots
\end{aligned}$$

ARCSIN AND ARCCOS

Just as you can construct a table of sine values for regular intervals of the angle A, so you can draw up an inverse table, showing values of A for regular intervals of sin A. *arcsin A* means 'the angle whose sine is'.

sin A	arcsin A
0.50	30°
0.51	30.66°
0.52	31.33°
0.53	32.01°
0.54	32.68°
0.55	33.37°

For a right-angled triangle whose sides are known, it is possible to use such a table to find the angle $A = \arcsin(a/c)$, where a is the side opposite the angle A and c is the hypotenuse.

Exercise 12

1 Using a pocket calculator or otherwise, find the following.
a $\sin(190°)$. **b** $\sin(0.5 \text{ radians})$. **c** $\cos(-\pi/4 \text{ radians})$.

2 By plotting points on graph paper, draw a sine wave, showing several cycles. *Note*: to draw the curve so that the amplitude is in proportion to the wavelength, use radian units.

3 On the same graph, sketch $\sin A$ and $\sin 2A$.

4 On the same graph, sketch $\sin A$ and $2 \sin A$.

5 In Figure 4.4E the radius of the circle equals 1. Find which line segments have the following lengths.
a $\sin A$. **b** $\cos A$. **c** $\tan A$. **d** $\operatorname{cosec} A$.
e $\sec A$. **f** $\cot A$.

Note:

$$\operatorname{cosec} A = \frac{1}{\sin A}$$

$$\sec A = \frac{1}{\cos A}$$

$$\cot A = \frac{1}{\tan A}$$

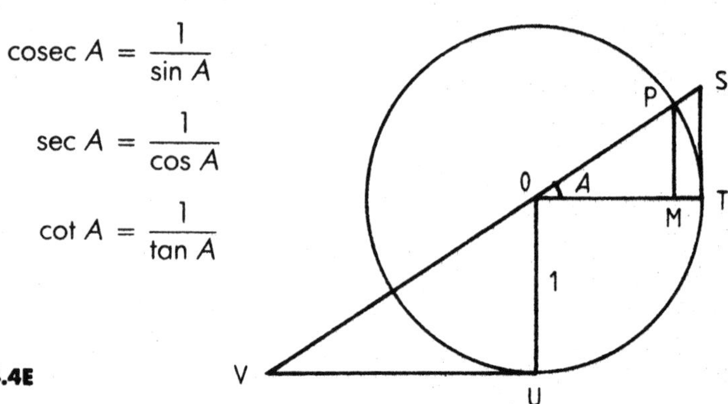

Figure 4.4E

6 If $\sin A = x$, find the following in terms of x.
a $\cos A$. **b** $\tan A$. **c** $\operatorname{cosec} A$. **d** $\sec A$.
e $\cot A$.

7 Find the sides of a right-angled triangle whose hypotenuse is 1 and whose other angles are 30° and 60°.

8 Find the height of an equilateral triangle with side length equal to 1.

9 Find the sides of a right-angled triangle whose hypotenuse is of length 3 and whose angles are 15° and 75°.

10 Find the height of a cathedral spire which subtends an angle of elevation of 29° at a distance of 400 m from its base on level ground.

11 Find the side lengths of the following regular polygons inscribed in a circle of unit radius.
a Pentagon. **b** Heptagon. **c** Octagon.
d Decagon. **e** Square. **f** Hexagon. **g** Triangle.

12 Using a pocket calculator or otherwise, find the following.
a $\arcsin(0.66)$. **b** $\arccos(0.66)$. **c** $\arcsin(-0.35)$.
d $\arccos(1.5)$.

13 Find the other two angles in the following right-angled triangles.
a $a = 5, b = 12, c = 13$.
b One side equal to 1 and hypotenuse equal to 2.
c Two shorter sides of 1 and 2 respectively.

14 Find the angle between the diagonal and an edge of a unit cube.

15 In a square-based pyramid whose sides are all 1 unit long, find the following angles.

a The angle between a sloping edge and the base plane.
b The angle between a sloping edge and the vertical.
c The angle between a sloping face and the base plane.
d The angle between a sloping face and the vertical.

16 For any triangle ABC prove the *sine rule*: $b \sin A = a \sin B$.

17 For any triangle ABC prove the *cosine rule*: $a^2 = b^2 + c^2 - 2bc \cos A$.

HOW THE SINE AND COSINE TABLES WERE COMPUTED

Hipparchus and Ptolemy were responsible for calculating the first tables of sines and cosines. How did they do this? A start can be made using Pythagoras' theorem to compute the sines and cosines of certain special angles, namely

$$\sin(45°) = \tfrac{1}{2}\sqrt{2} = \cos(45°)$$
$$\sin(60°) = \tfrac{1}{2}\sqrt{3} = \cos(30°)$$
$$\sin(30°) = \tfrac{1}{2} = \cos(60°)$$

You may also recall from Section 1.2 that Euclid was able to show that the side of a decagon inscribed in a unit circle equals $1/\tau$ so that, in other words

$$\text{chord}(36°) = \tfrac{1}{2}\sqrt{5} - \tfrac{1}{2}$$

This gives

$$\sin(18°) = \tfrac{1}{4}\sqrt{5} - \tfrac{1}{4} = \cos(72°)$$

So far, so good. To proceed further and to find sines and cosines of any other angle, it was necessary to have formulae equivalent to the following:

$$\cos(A + B) = \cos A \cos B - \sin A \sin B$$

giving

$$\cos(2A) = \cos^2 A - \sin^2 A = 2\cos^2 A - 1$$

and

$$\cos A = \sqrt{[\tfrac{1}{2}\{\cos(2A) + 1\}]}$$

With the help of these formulae it was possible to bisect systematically the starting angles of 45°, 60°, 30° and 18° in an iterative manner to obtain angles as small as seemed practically necessary, and to add the parts in any combinations of pairs to obtain all angles between 0° and 90°.

Of course, this was a monumental task of calculation, involving square root after square root, and any error would be passed on to all subsequent calculations. Once done, however, it was good for all time. The tables of sines and cosines, like the calculation of π, were triumphs of numerical methods in the mathematics of the ancient Greeks, based upon the repeated use of Pythagoras' theorem.

LATER DISCOVERIES

With the coming of the age of calculus in seventeenth-century Europe, new methods were discovered for computing the values of sines and cosines. To cut a long story short, and assuming that the angle A is measured in radians,

$$\cos A = 1 - \frac{A^2}{2!} + \frac{A^4}{4!} - \frac{A^6}{6!} + \ldots$$

$$\sin A = A - \frac{A^3}{3!} + \frac{A^5}{5!} - \frac{A^7}{7!} + \ldots$$

Both of these sums (*series*) have infinitely many terms but, as the factorial denominators soon grow very large indeed, it is in practice necessary to compute only a relatively small number of terms. This method is, of course, far more efficient than the Greek method and is essentially what your calculator or computer performs each time that you press the sine or cosine button.

(*Note*: the lengthy calculations of square roots and sines and cosines which we have been looking at in Chapter 4 should give you a firm impression of why they are relatively time consuming even for fast computers; so, if a computer program that you have written is slow running, this may be due to repeated computations of sines, cosines and square roots.)

Exercise 13

1 Using Figure 4.4F and looking for similar triangles, prove the following.

a $\cos(A + B) = \cos A \cos B - \sin A \sin B$.
b $\sin(A + B) = \sin A \cos B + \sin B \cos A$.

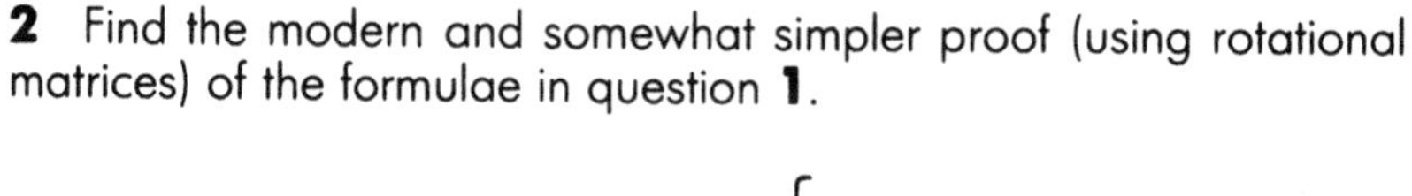

2 Find the modern and somewhat simpler proof (using rotational matrices) of the formulae in question **1**.

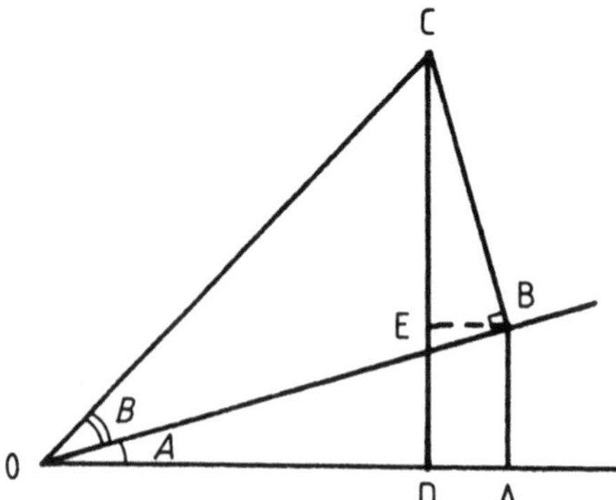

Figure 4.4F

3 Taking $A = 1$, give the first term in each of the above series expansions of sin A and cos A which is less than 0.000001 (a millionth).

5 Computer Drawings

Computer graphics includes, of course, the possibilities of colouring and animation, but this book is only concerned with drawing. We shall draw only straight lines from point to point. This is known as *vector graphics*. Suitable output devices include a high-resolution visual display unit (VDU) screen, a dot-matrix printer with screen-dump commands or a plotter, as used here.

You may already know a little school mathematics and have learnt to write a few programs. Whether or not this is so, programming graphics is an excellent way to teach yourself either mathematics or programming, because it supplies the programmer with immediate visual feedback:

you think → you write → it draws → you look → you think → ...

PROGRAM LANGUAGE

Code given in this book is in a specially simplified version of BASIC so as to be easily recognised by readers using any dialect of BASIC. Indeed, users of any programming language should find it easy to follow; my aim is to discuss the mathematics of the drawings and not programming. The job of putting the program pieces together into completed programs and adapting them for specific machines is left up to you.

Most of the programs used in this book are very simple, with much use being made of the FOR...NEXT... loop. Here is a quick guide to the BASIC terms to be used; they are nearly all familiar mathematical functions.

BASIC	Command Effect
The FOR...NEXT...loop	To repeat the sequence of commands between FOR and NEXT a specified number of times
The IF...THEN...gate	Executes the THEN... command if the IF... statement is true
GOSUB...RETURN	Executes a subroutine
DIM X(1000)	Sets up an array named X with 1000 elements given an initial value of 0

DRAWING COORDINATES

I shall use two coordinate systems (Figure 5.0A), *Cartesian* (X,Y) and *polar* (R,A), and assume that they are laid out on the drawing board with the origin at its centre as shown below. Readers using origins placed

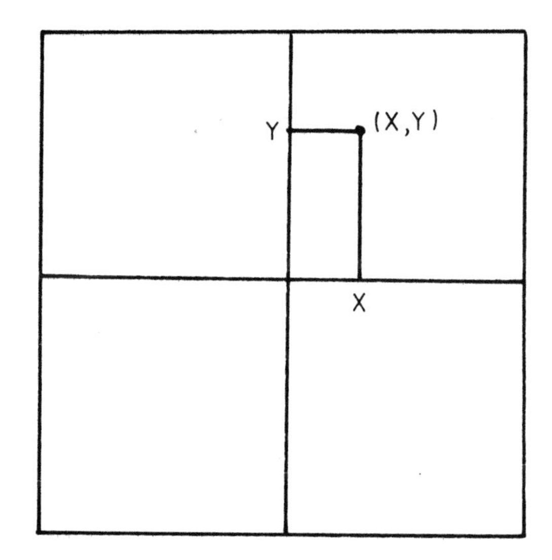

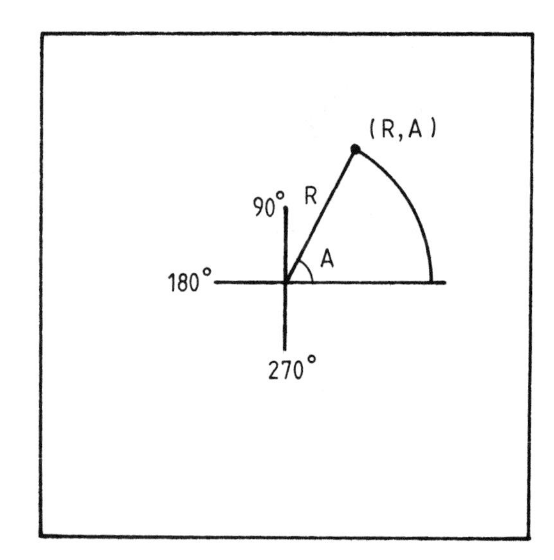

Figure 5.0A

elsewhere and different scales will find it an easy matter to translate and scale their drawings accordingly.

MOVE AND DRAW

At the beginning of each program the *present location* of the pen is (0,0). The commands for moving the pen to a new present location are MOVE X,Y and DRAW X,Y. The latter draws a straight line from the old present location to the new present location; the former does not. Here, by way of example, is the routine for drawing a straight line from (20,30) to (60,−80):

```
MOVE 20,30
DRAW 60,-80
```

MOVEP AND DRAWP

MOVEP and DRAWP perform the same tasks as MOVE and DRAW, the only difference being that they refer to polar coordinates instead of Cartesian coordinates.

CONVERTING POLAR COORDINATES TO CARTESIAN COORDINATES

The polar coordinates (R,A) can be converted to Cartesian coordinates (X,Y) as follows

```
X=R*COS(A)
Y=R*SIN(A)
```

DEGREES AND RADIANS: THE CONVERSION FORMULAE

$$360 \text{ degrees} = 2\pi \text{ radians}$$

$$1 \text{ degree} = \frac{2\pi}{360} \text{ radians} \qquad 1 \text{ radian} = \frac{360}{2\pi} \text{ degrees}$$

$$X \text{ degrees} = \frac{2\pi X}{360} = \text{radians} \qquad Y \text{ radians} = \frac{360Y}{2\pi} = \text{degrees}$$

BASIC MATHEMATICAL FUNCTIONS

BASIC	Meaning
+	Add (+)
−	Subtract (−)
*	Multiply (×)
/	Divide (÷)
X ↑ 3	x to the power 3 (x^3)
SQR	Square root ($\sqrt{}$)
SIN, COS and TAN	sine (sin), cosine (cos) and tangent (tan)
ASN, ACS and ATN	arcsin, arccos and arctan
LOG and LN	logarithm to base 10 (log) and logarithm to base e = 2.718... (ln)
EXP(X)	e^x
ABS(X)	The positive value of x ($\|x\|$)
INT(X)	x rounded down to its nearest integer value
RND	Returns a value between 0 and 1 to simulate the continuous uniform distribution of a random variable

Exercise 14

1 Find the numerical value in degrees of the following angles.
a 1 radian. **b** 3.141 59 radians. **c** 6 radians.
d $\pi/2$ radians. **e** $\pi/4$ radians. **f** $\pi/3$ radians.
g $2\pi/5$ radians. **h** $\pi/5$ radians.

2 Find the numerical values in radians of the following angles.
a 360°. **b** 1°. **c** 10°. **d** 90°. **e** 180°. **f** 60°.
g 45°. **h** 30°.

3 Give the formulae for converting Cartesian coordinates into polar coordinates.

4 What shapes are drawn by the following?

a
```
MOVE 0,0
DRAW 50,0
DRAW 50,50
DRAW 0,50
DRAW 0,0
```

b
```
MOVEP 50,0
DRAWP 50,72
DRAWP 50,144
DRAWP 50,216
DRAWP 50,288
DRAWP 50,0
```

5 Give the meaning of ACS or arccos in words.

37 Touching stars in concentric rings

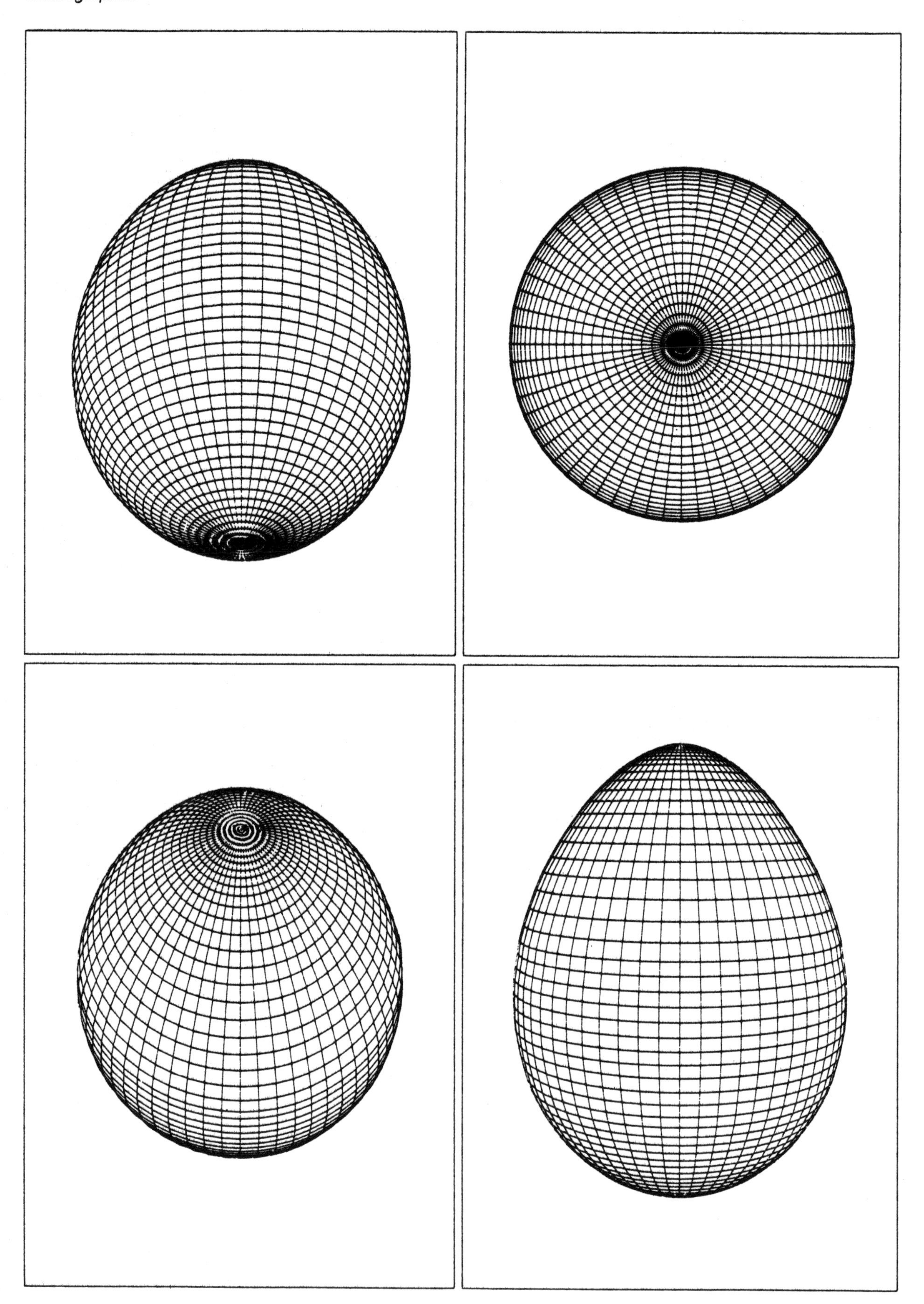

38 Four views of an egg

39 Venus

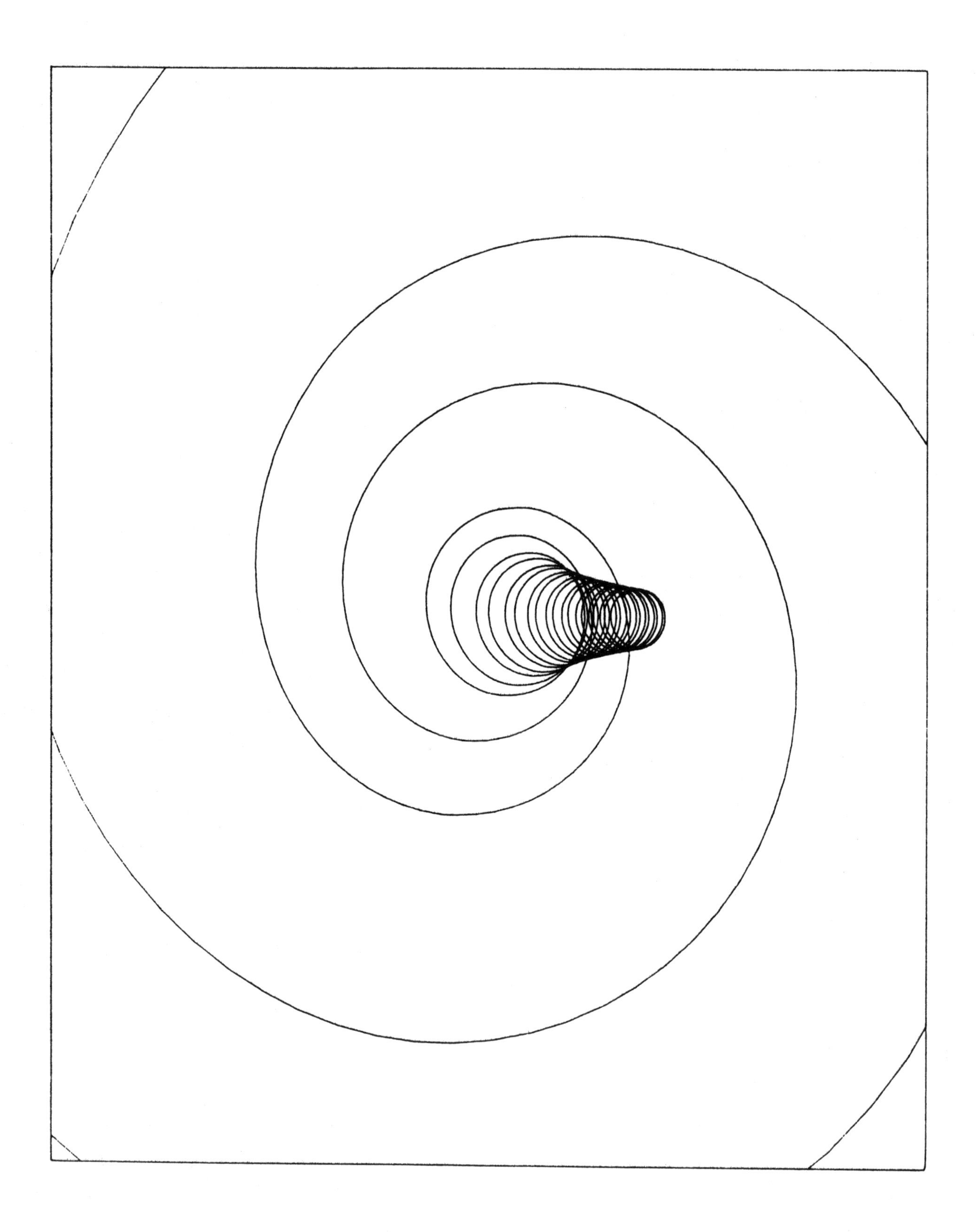

40 Plughole vortex

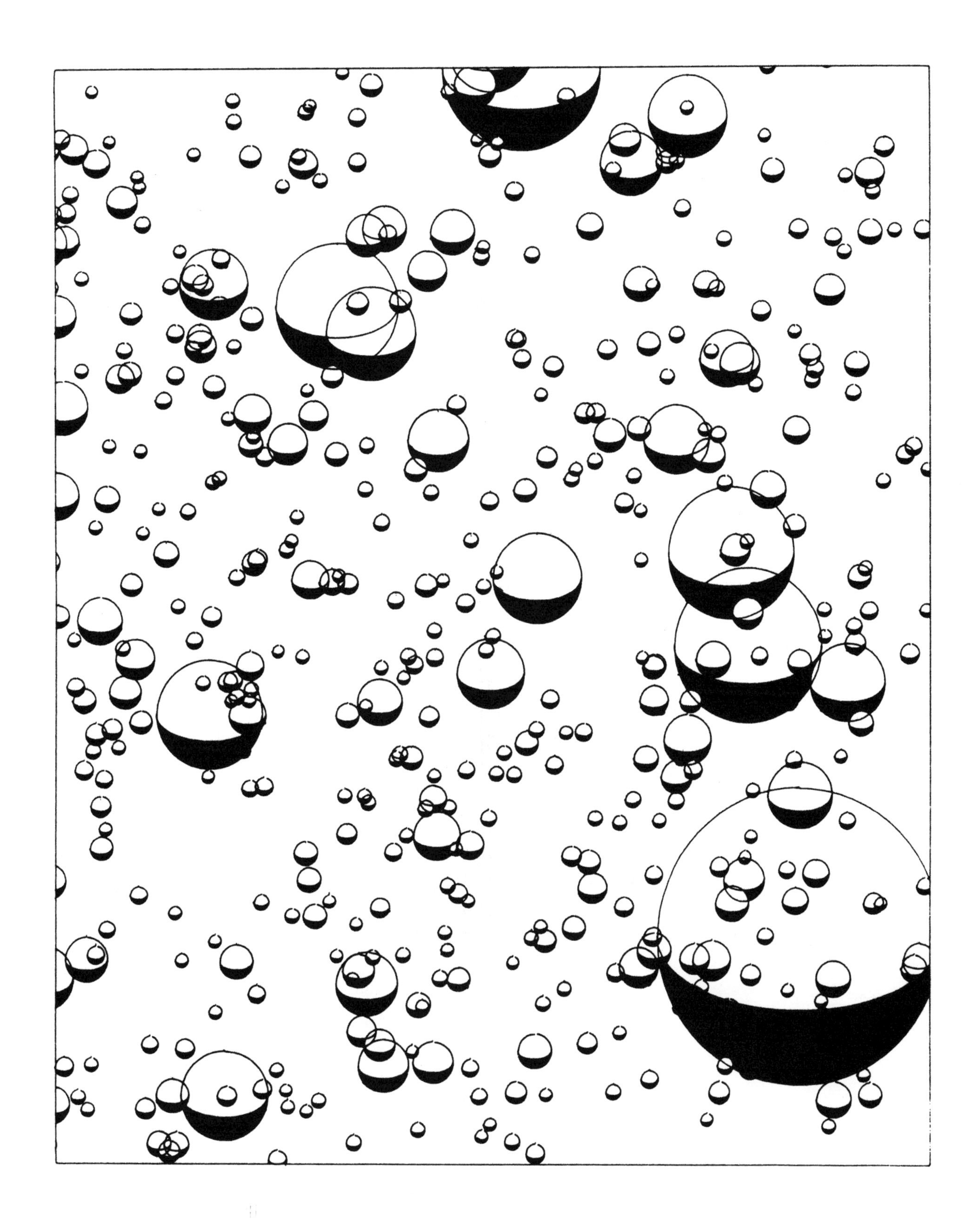

41 Bubbles

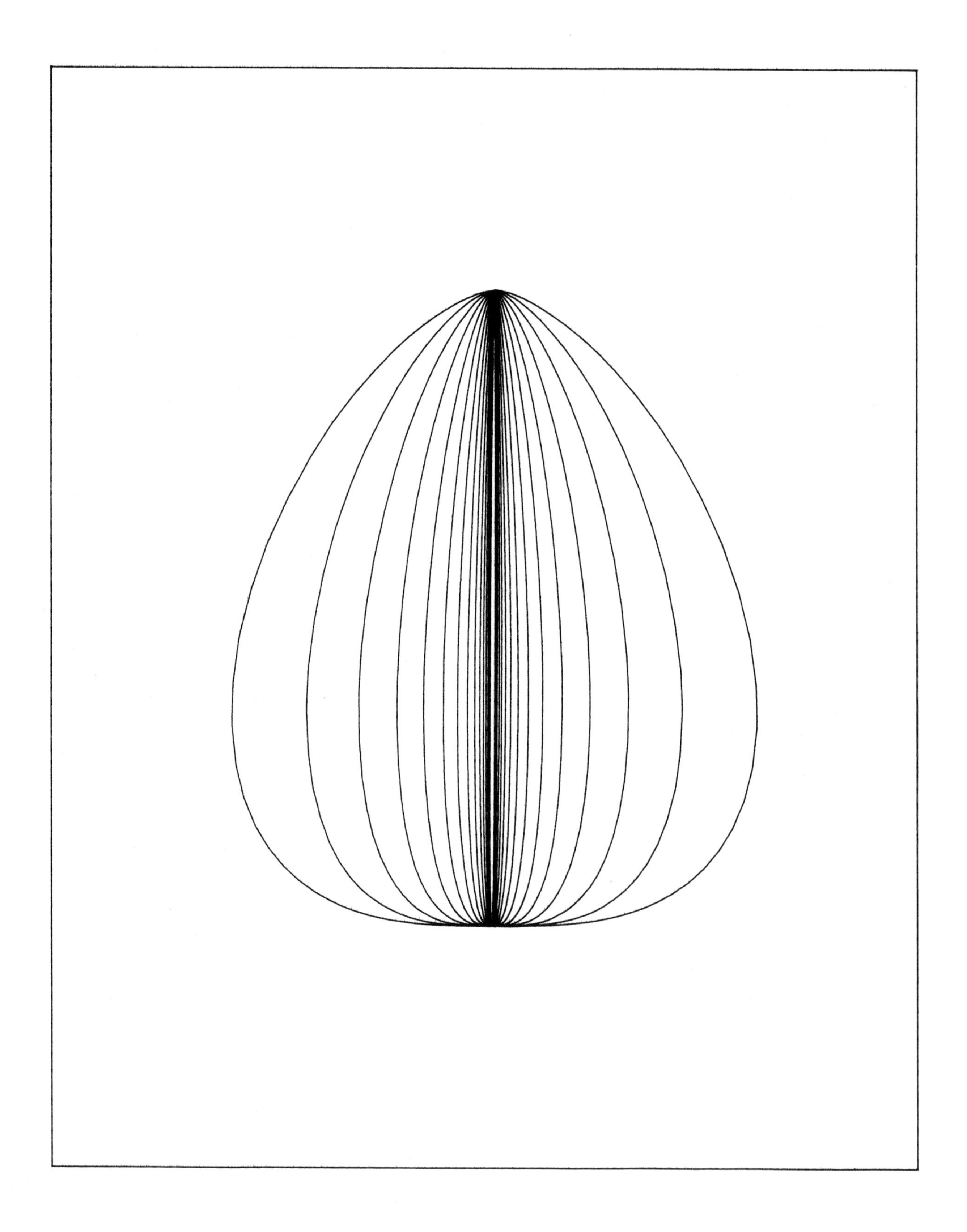

42 Curves of collineation (after Lawrence Edwards)

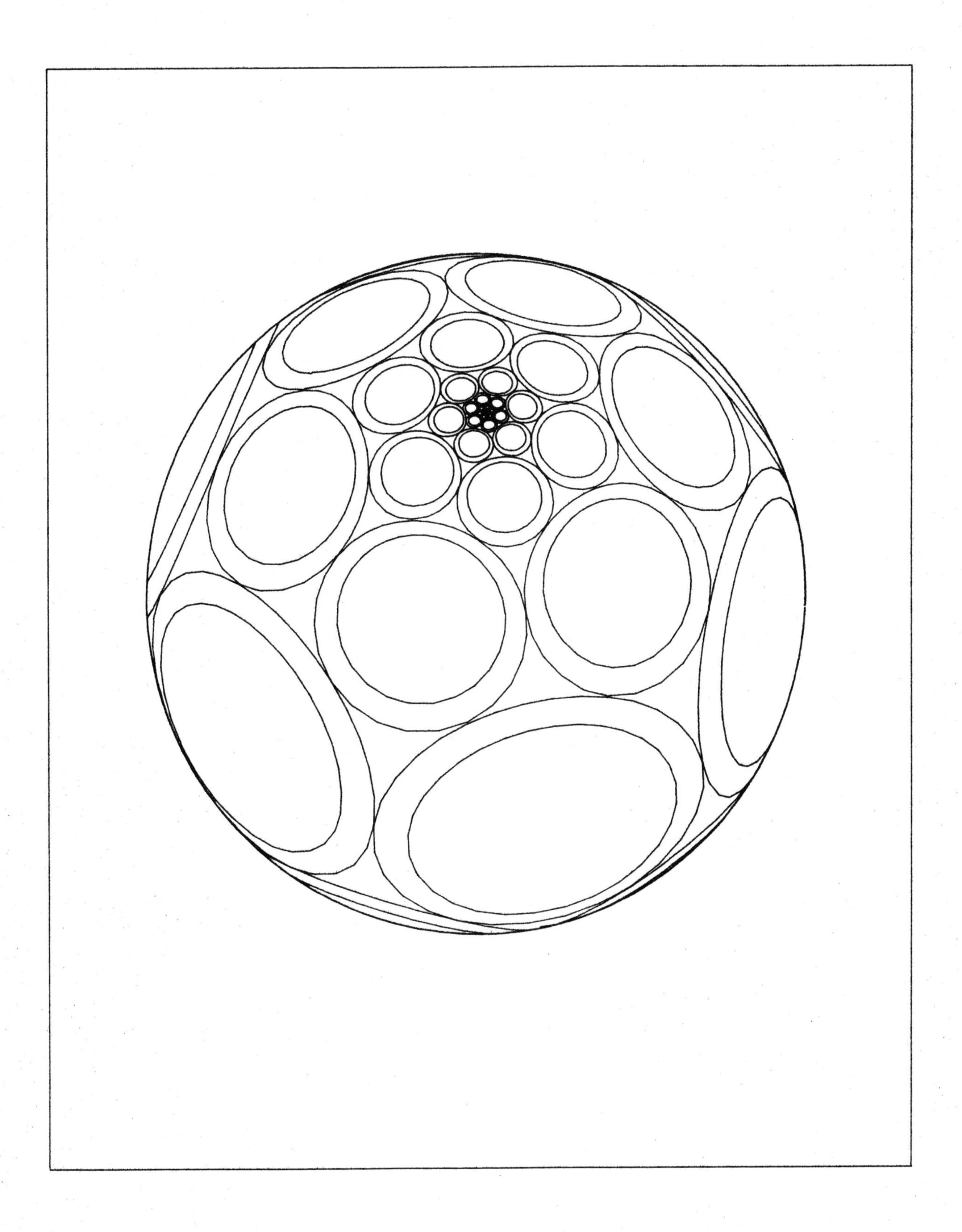

43 Sphere

44 Crinkles

45 Waves

46 Ocean view

47 The complete Fermat spiral: $r = \pm\sqrt{a}$

5.1 On drawing a daisy

In this section we learn how to draw the following simple forms:

- regular polygons;
- the circle;
- various Cartesian curves;
- various polar curves, including three types of spiral;
- a daisy.

CARTESIAN CURVES

The FOR...NEXT... loop enables us to repeat a line-drawing routine any specified number of times, as in the following example of a graph-drawing routine referring to Cartesian coordinates!

```
FOR X=-100 TO 100
Y=X↑2
IF X=-100 THEN MOVE X,Y
DRAW X,Y
NEXT
```

SCALING AND TRANSLATING

The following example shows how to magnify a small part of the above curve and move it down the drawing board. Scaling and translating of a drawing are achieved by multiplying and adding respectively the (X,Y) coordinates.

```
FOR X=-10 TO 10 STEP 0.25
Y=X↑2
X=10*X
Y=Y-80
IF X=-100 THEN MOVE X,Y
DRAW X,Y
NEXT
```

LOOPS WITHIN LOOPS

Putting the above curve-drawing loop inside another loop enables us to draw several curves on the same graph. For example, here is the routine for drawing the *family* of curves $y = x$, $y = x^2$, $y = x^3$, ..., $y = x^{10}$:

```
FOR N=1 TO 10
FOR X=-10 TO 10
Y=X↑N
:
:
NEXT X
NEXT N
```

Exercise 15

Write programs to draw the following Cartesian curves, together with the two axes.

1 Powers of x (Drawing **48**).
a $y = x^2$ for $x = -3$ to 3.
b $y = x^n$ for $n = 0, 1, 2, 3, 4$ and 5 (loop within a loop, giving six curves).
c $y = x^n$ for $n = 1.0, 1.2, 1.4, \ldots, 2.0$ and $x = 0$ to 3 (six curves).
d $y = x^n$ for $n = 0, 0.2, 0.4, \ldots, 2.0$ and $x = 0$ to 3 (11 curves).
e $y = x^{-1}$ for $x = 0.1$ to 10.
Note that your program will crash if at any time you attempt to compute x^n when x is zero and n is negative, or when x is negative and n is not an integer.

2 Waves (Drawing **49**) It is assumed that your computer computes SIN(X), etc., where X is in radians. If you need to convert into degrees use X=180*X/3.14159. If you need to scale your drawing up, use the same magnification value for both the x and the y coordinates.
a $y = \sin x$, for $x = -10$ to 10.
b $y = \cos x$, for $x = -10$ to 10.
c $y = n \sin x$, for $x = -10$ to 10 and $n = 1, 2, 3$ and 4 (four curves).
d $y = \sin(nx)$, for $x = -10$ to 10 and $n = 1, 2, 3$ and 4 (four curves).
e $y = \sin(x) + \sin(x + 1.57)$, for $x = -20$ to 20.
f $y = \sin(2x) + \sin(3x)$, for $x = -10$ to 10 (a compound wave).
g $y = \sin(x) + \sin(2x) + \sin(3x)$ (a compound wave).

3 Exponential curves.
a $y = 2^x$, for $x = -4$ to 4.
b $y = n^x$, for $n = 0.5, 1.0, 1.5, \ldots, 3.0$ and $x = -4$ to 4 (six curves).

4 Logistic curves.
a $y = a/(b + c^{-x})$, experimenting with various values of a, b and c.

The exponential curve gives a good approximation to the curve of *growth* (e.g. the weight of a daisy, the human population) in the early stages. Over the longer term, however, the logistic curve (Figure 5.1A) gives a better picture, as it assumes an upper limit to growth. It is based upon the assumption that the rate of growth is proportional to the difference between the present size and the upper limit. Interestingly enough, the same *sigmoidal* shape of curve is given by $y = \arctan(a^x)$.

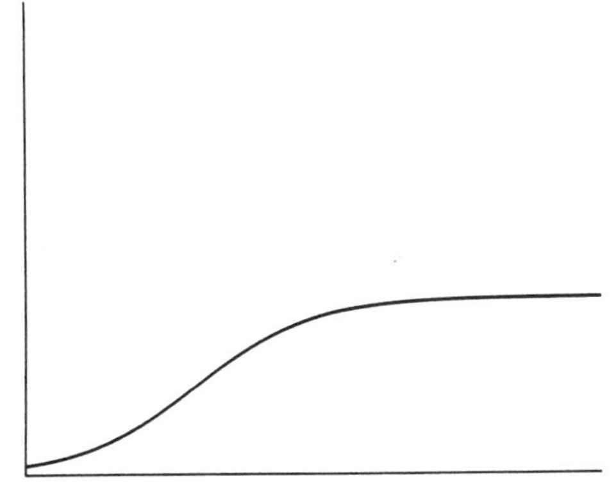

Figure 5.1A

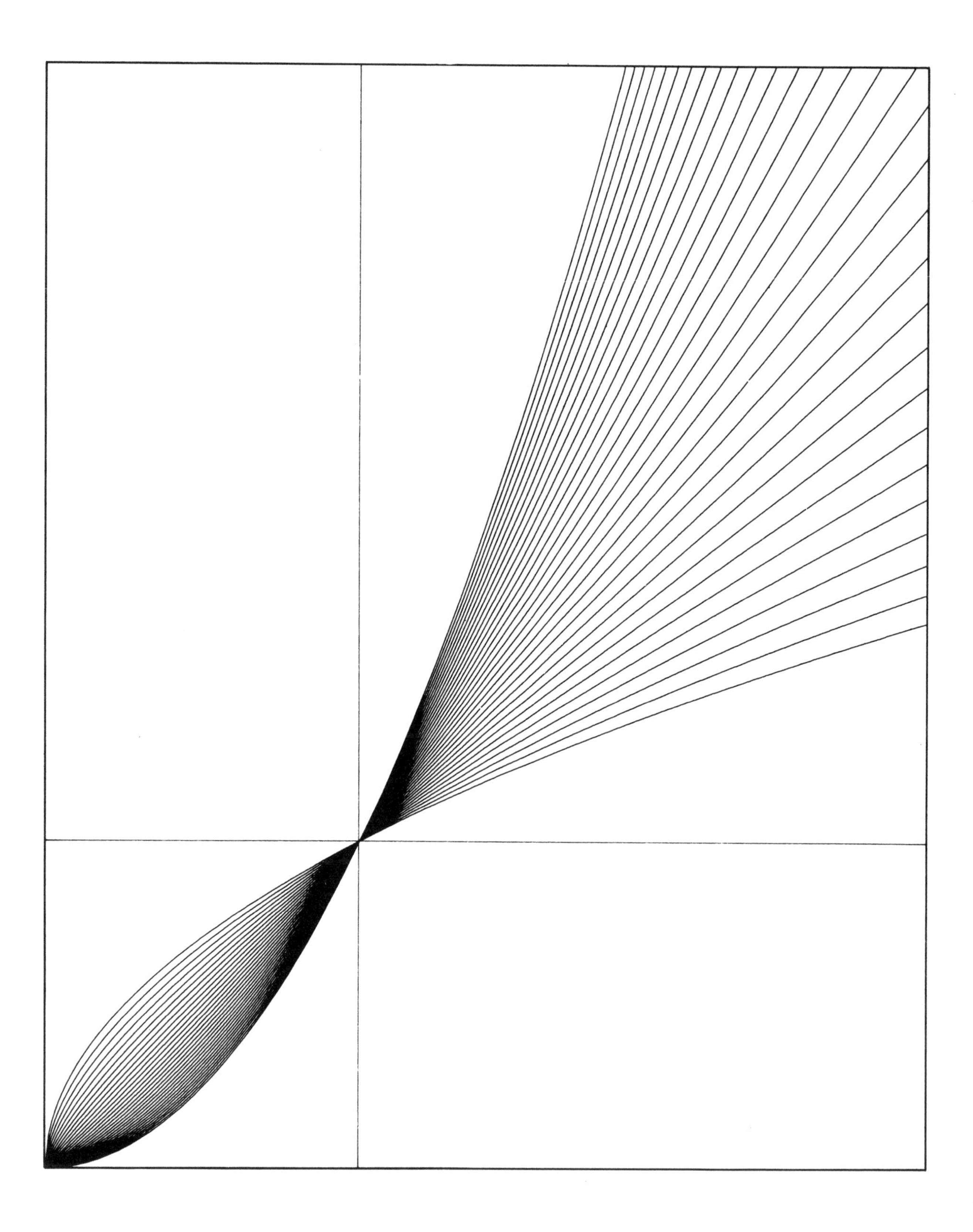

48 Powers of x

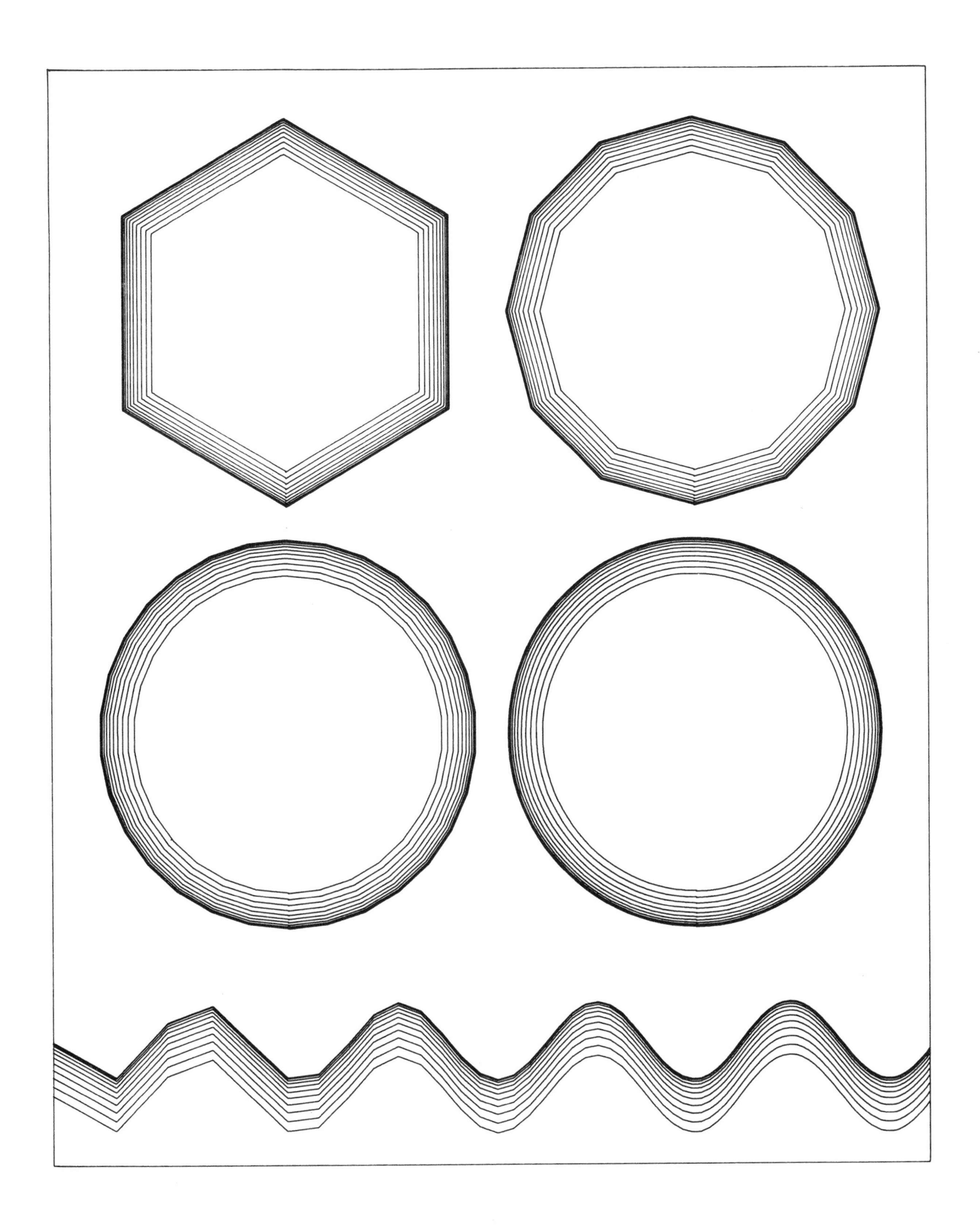

49 Polygons, a circle and a sine wave

REGULAR POLYGONS AND CIRCLES

When you draw regular polygons and circles, it is more convenient to work in polar coordinates (Drawing **49**). To draw a regular octagon, for example, divide a circle into eight equal parts and join these points with straight lines.

```
FOR A=0 TO 360 STEP 360/8
IF A=0 THEN MOVEP 50,A
DRAWP 50,A
NEXT
```

To draw a circle, replace 8 in the above program by 30 or 60.

STAR POLYGONS

```
INPUT N
INPUT M
FOR S=0 TO N
A=360*S*M/N
IF S=0 THEN MOVEP 50,A
DRAWP 50,A
NEXT
```

In the above program, N is the number of points in the star. Choose M to be a whole number between 1 and N which shares no common factor with N. (What happens if M and N have a common factor?)

SPIRALS

A *spiral* is a plane curve traced out by a point which winds about its pole with continually increasing radius. A *monotonic* function $R=F(A)$ is any function for which increases in A correspond to increases in R. So, to draw a spiral with, for example, five turns, use

```
FOR S=1 TO 600
R=F(A)          (where F(A) is some specified monotonic function)
IF S=1 THEN MOVEP R,A
DRAWP R,A
A=A+3                                (A is measured in degrees)
NEXT
```

(*Note*: the phrases in parentheses are only explanatory and are not part of the program.) The increments by which A is increased at each time must be small enough to give the effect of a smoothly flowing curve. Here are three such spirals.

a Archimedes' spiral (Drawing **50**)

R=K*A (where K is some constant)

b The equiangular spiral (Drawing **51**)

R=K↑A

c Fermat's spiral (Drawing **52**)

R=SQR(A)*K

Spiral **a** (Figure 5.1B) increases its radius by the same amount on every turn (see p. 77). Spiral **b** is called the equiangular spiral because its path

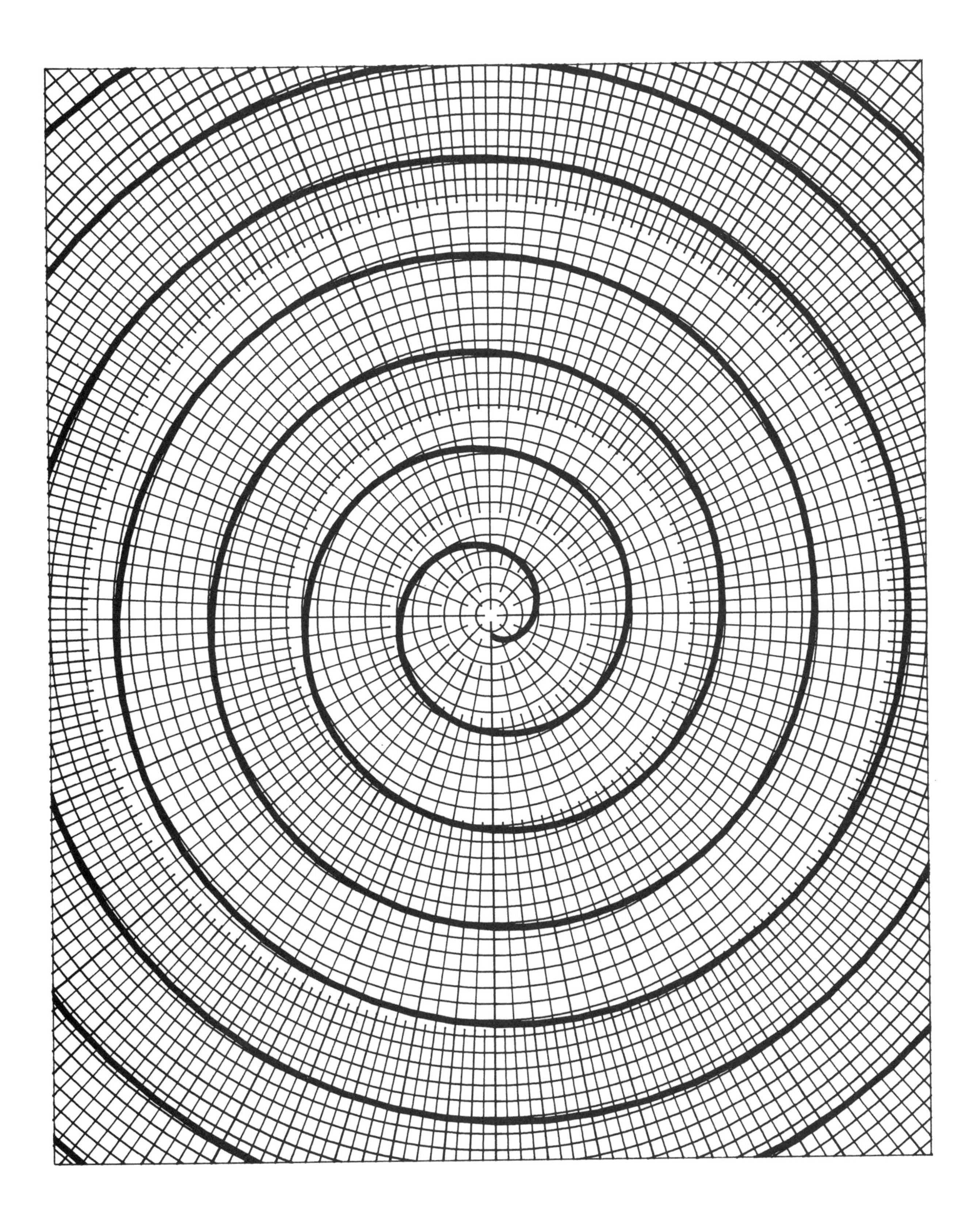

50 Archimedes' spiral: $r = ka$

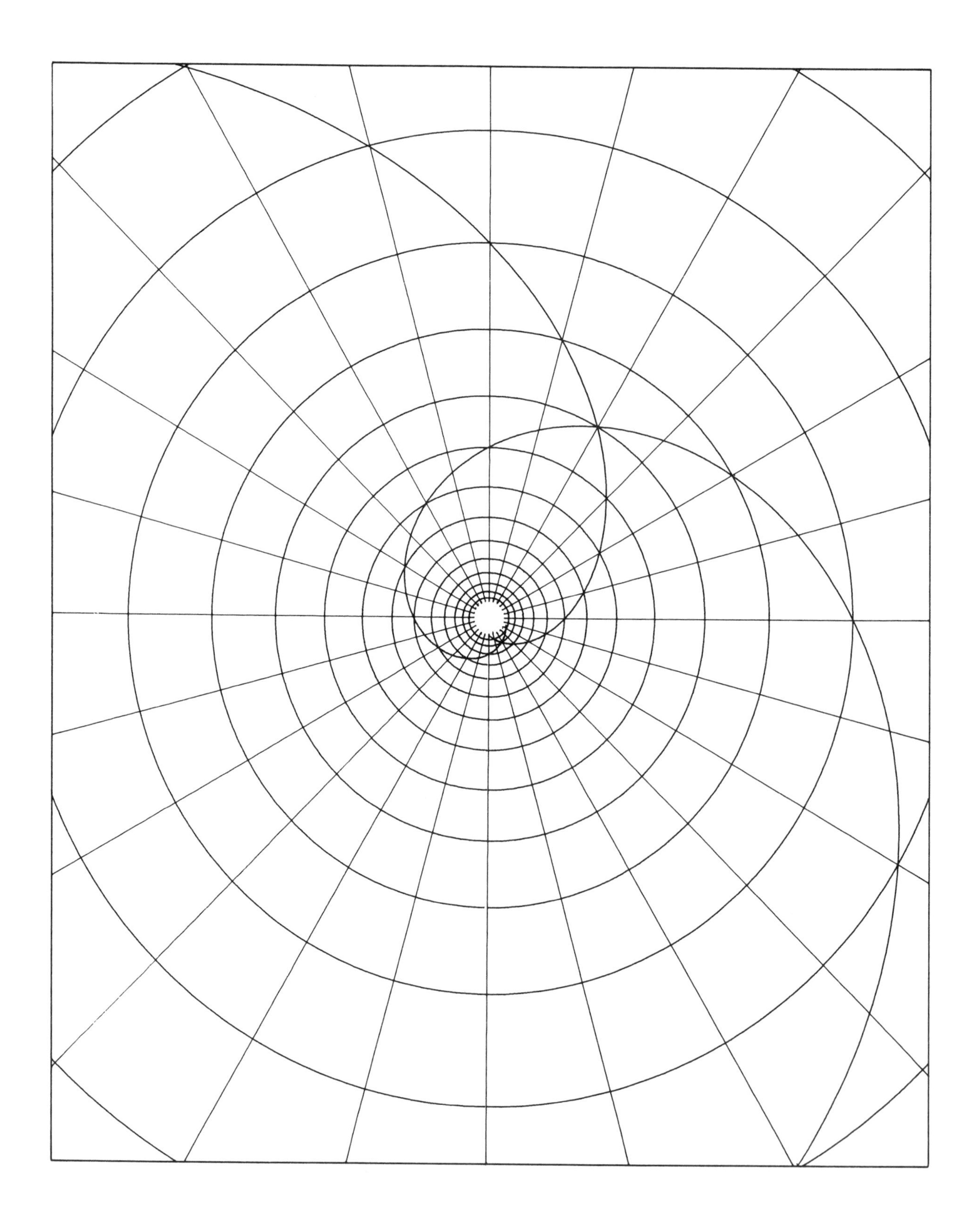

51 Equiangular spiral: $r = k^a$

52 Fermat's spiral: $r = \sqrt{(ka)}$

Figure 5.1B

Figure 5.1C

Figure 5.1D

continues to make a constant angle with the radius. It is also called the logarithmic spiral, and each turn brings about a proportional increase in radius. Sea shells and snail shells display this shape very clearly (Figure 5.1C). Spiral **c** (Figure 5.1D) has the property of enclosing equal areas with every turn, and we shall use this property to construct a daisy.

DAISIES

The following program draws polygons centred at regular angular intervals of D, called the *divergence*, on Fermat's spiral:

```
INPUT K
INPUT C
D=360/C
FOR S=1 TO 100
R=K*SQR(A)
FOR...
(loop for drawing polygons centred at (R,A))
NEXT
A=A+D
NEXT S
```

The polygon-drawing loop should resemble the following:

```
FOR T=0 TO 6
AA=360*T/6 + A
RR=5
X=RR*COS(AA)+R*COS(A)
Y=RR*SIN(AA)+R*SIN(A)
IF T=0 THEN MOVE X,Y
DRAW X,Y
NEXT T
```

The choice of K determines how tightly wound is the spiral (not actually drawn) while the choice of C determines the divergence. A true daisy has a divergence of $360°/\tau = 222.49°$. Experiment with a variety of values for C, and hence for D, to see the effects. Each choice gives a different pattern of *secondary* spirals which form an interlocking system, with so many winding clockwise and so many winding anticlockwise. $C = \tau = 1.618\ 034$ is the only choice which gives an even packing of polygons. The secondary spirals in this case occur in *Fibonacci numbers*: 1, 2, 3, 5, 8, 13, 21, 34, 55, ...

Plants in the real world (such as daisies, sunflowers, pineapples, pinecones and many more) demonstrate this pattern clearly (Drawings **53–58**). It arises naturally as a space-saving device.

53 Cylindrical phyllotaxis

54 Cylindrical phyllotaxis, with vertical compression

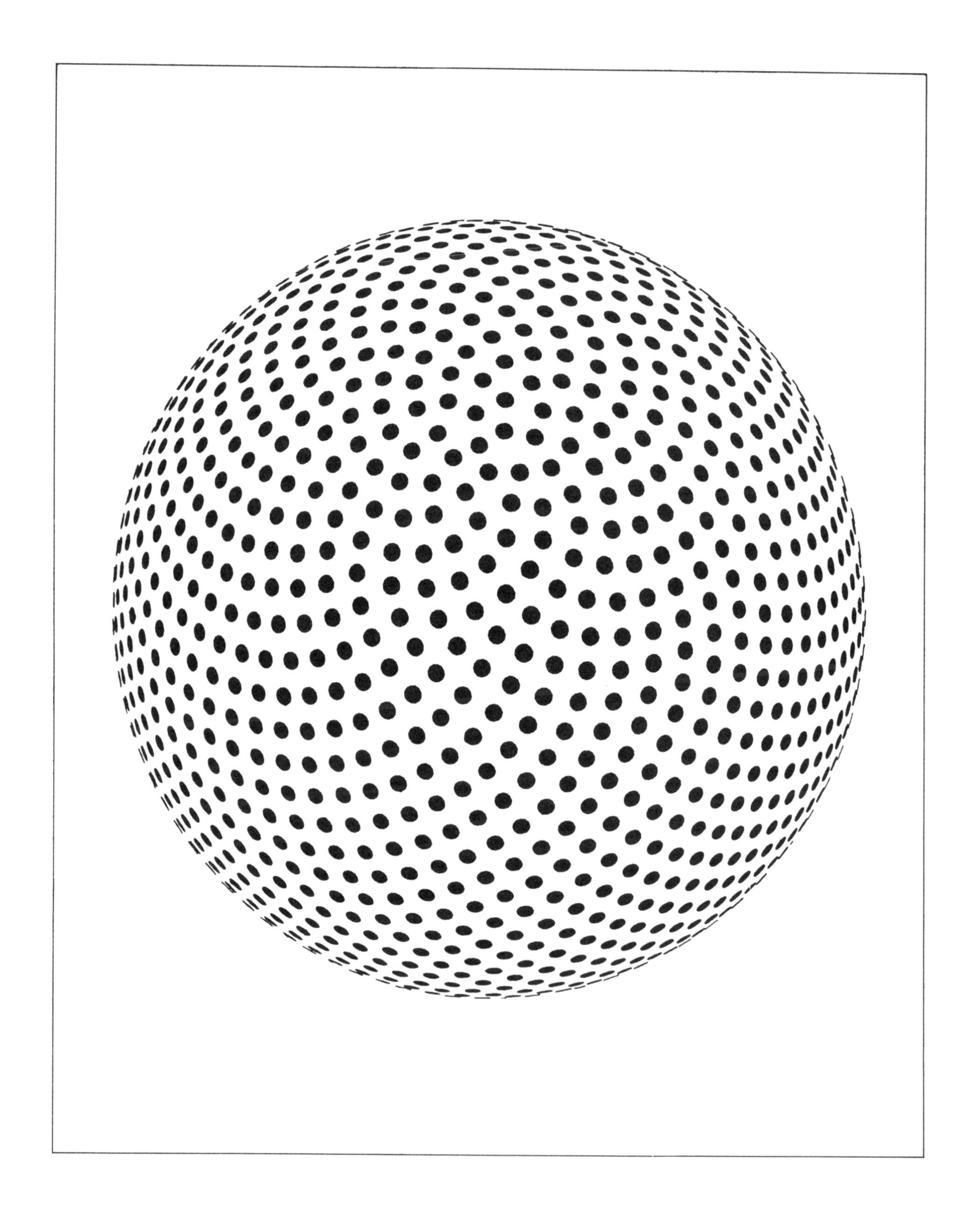

55 Spherical phyllotaxis

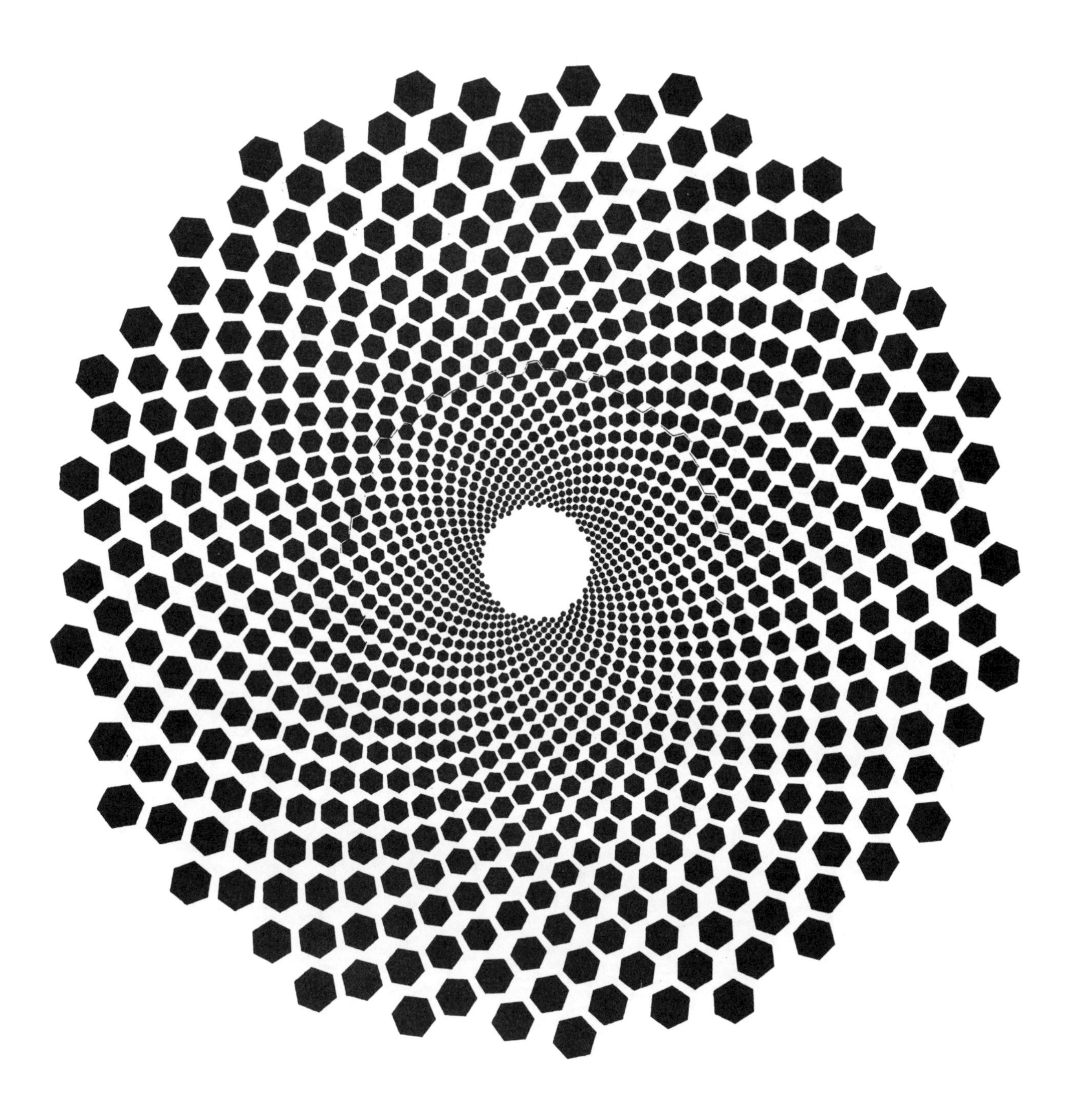

56 Exponential growth

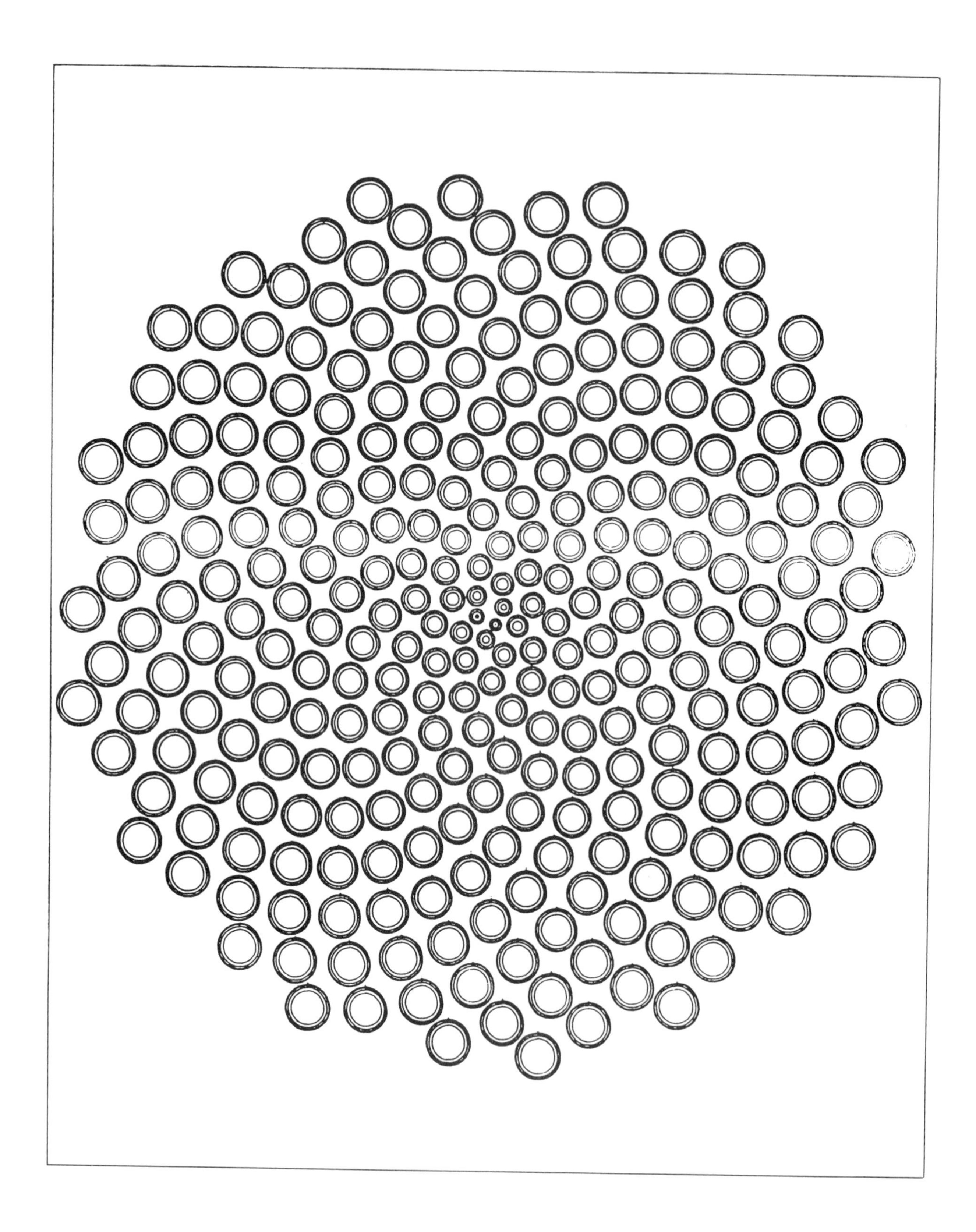

57 Logistic growth

58 Sunflower

Exercise 16

1 Draw the following spirals.

a *Archimides' spiral.* R=K*A, for A=0 to 3600. Try several values of K.

b *Fermat's spiral.* R=K*SQR(A), for A=0 to 3600.

c *The complete Fermat spiral.* This is the set of points (R,A) counting both solutions to the equation $r^2 = k^2a$, for a=0 to 3600:

```
FOR A=−3600 TO 3600
IF A ⩾ 0 THEN R=K*SQR(A)
IF A < 0 THEN A=−A; R=K*SQR(A): R=−R
```

d *The hyperbolic spiral.* R=K/A, for A=30 to 7200. Try, for example, K=7200. This spiral corresponds to a perspective view down the axis of a regular helix.

e *The lituus (Bishop's crook).* R=K/SQR(A), for A=30 to 7200. Try K=100.

f *The logarithmic or equiangular spiral.* R=K↑A, for A=0 to 3600. Experiment widely with values of K.

2 *False daisies* (Drawings **59** and **60**). Using the program listing on p. 126 draw the following false daisies. For each drawing, ensure that the polygons are small enough to get a hundred or so within the picture frame, and adjust K (spiral tightness) until the polygons are nearly touching.

a $C = \sqrt{2}$. **b** $C = e = 2.718$. **c** $C = \pi$.
d $C = \sqrt[3]{2}$. **e** $C = 5$. **f** 5/8.

3 *True daisy* (Drawing **61**). $C = 1.618\,034$.

4 Program a true daisy with polygons growing exponentially, RR=(1.01)↑S and R=K↑A.

5 Program a true daisy with polygons growing logistically RR=10/(1+(1.1)↑(S-50)) and R=R+K*SQR(RR), for S=0 to 100.

6 Program the following polar curves.

a *Rose curves* (Drawing **62**). R=80*SIN(M*A). Try M = 1, 2, 3, 4, 5, 6, 7, 10, 1/2, 1/3. Let A range from 0 to 360 for whole number values of M, and 0 to 360/M for the last two examples of fractional M. Odd M gives M petals; even M gives 2M petals.

b *Cardioid.* R=40*(1+SIN(A)), for A=0 to 360.

c *Freeth's nephroid.* R=25*(1+2*COS(A/2)), for A=0 to 720.

d *Cayley's sextic.* R=80*(SIN(A/3)↑3), for A=0 to 1080.

e *The cochleoid.* R=80*SIN(A)/A, for A=−1080 to 1080, except for R=80 when A=0.

f *Conchoid of Nicomedes.* R=10/SIN(A)+K, for A=3 to 177 and A=183 to 357. Try several values of K, such as K=70, 40, 20 and 10.

g *Cissoid of Diocles.* R=20*(COS(A)↑2)/SIN(A), for A=3 to 177.

7 *Conics* (focus at pole). R=L/(1+E*COS(A)), for $A=0$ to 360. L is the semilatus rectum; try $L=30$. E is the eccentricity of the conic. Try several values of E to obtain the following.

a The circle: $E=0$.
b Ellipses: $0 < E < 1$.
c The parabola: $E = 1$.
d Hyperbolae: $E > 1$.

General note: you may need to set a trap to discontinue a curve when R exceeds some large value.

Reading

Cook, 1914; Cundy and Rollett, 1961; Dixon, 1983a, 1983b, 1985a; Lawrence, 1972; Lockwood, 1961; Stevens, 1977.

59 False daisies: divergence of 360°/C, with $C = \pi$ and e

60 False daisies: $C = \sqrt{5}$ and 1.309

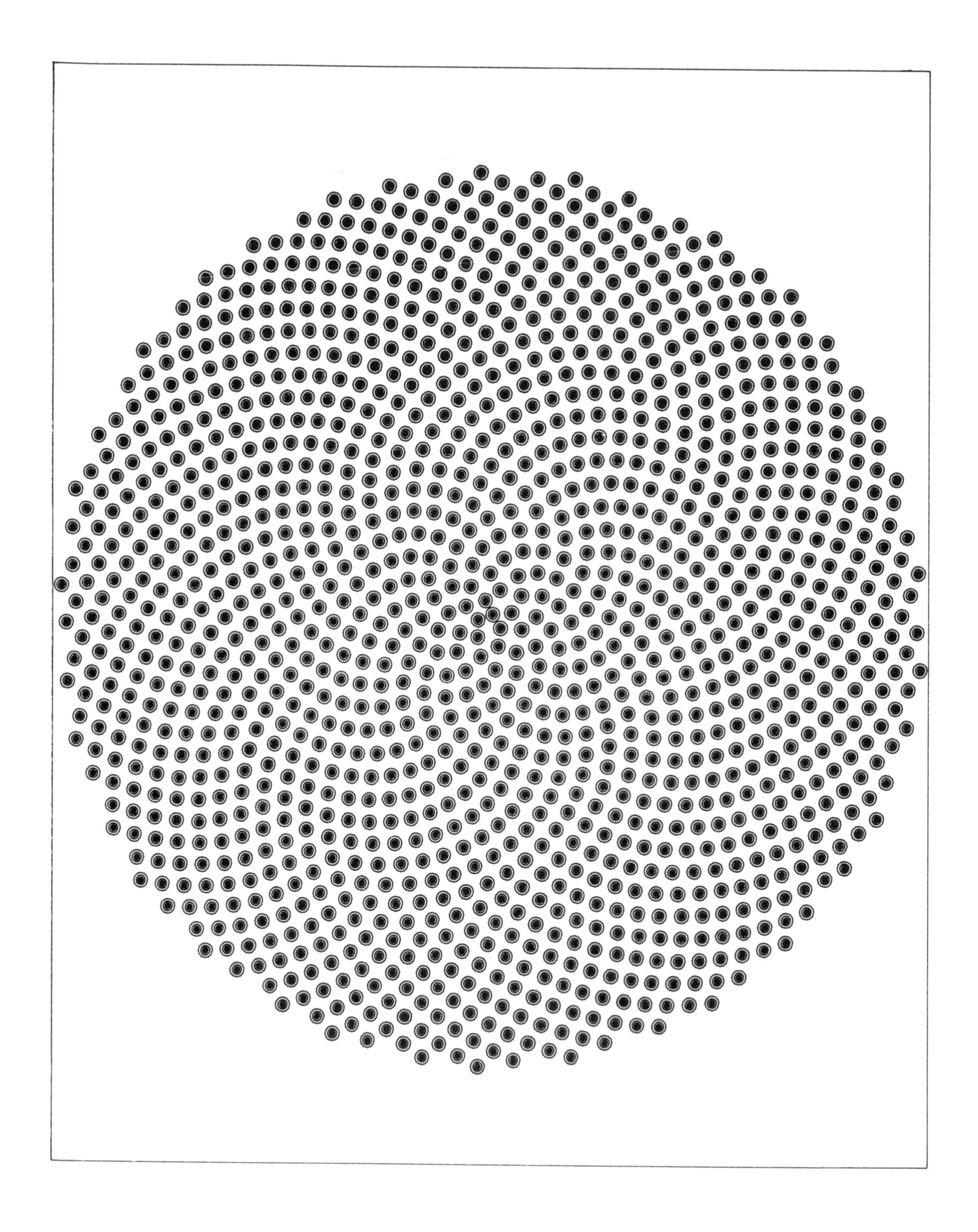

61 True daisy: $C = \tau$

62 Rose curves with a twist

5.2 Drawing in perspective

To represent three-dimensional objects start by giving each point a triple coordinate (x, y, z), and imagine that the xy plane is in the drawing plane, while the z axis points straight at you. Next you must decide where on the z axis to place your eye, the point of view for projecting the object onto the drawing plane (Drawings **63** and **64**).

PARALLEL PROJECTION

The simplest solution is to place your eye at infinity (with an infinitely powerful telescope, of course!). This case of projection is called *parallel* (and also *orthogonal*) projection (Figure 4.2A). It is used by architects and engineers a lot. There are no effects of diminution of size with receding distance. Parallel lines appear parallel. To draw in parallel projection, simply ignore the z coordinates and represent the point (x, y, z) at the point (x, y) on the drawing board.

POINT PROJECTION

Alternatively, you can choose to place your eye at a distance d from the drawing board on the z axis, and project from this viewpoint $(0, 0, d)$. This leads to the following conversion formula:

$$(x, y, z) \rightarrow (x_1, y_1)$$

where

$$x_1 = \frac{xd}{z - d} \text{ and } y_1 = \frac{yd}{z - d}$$

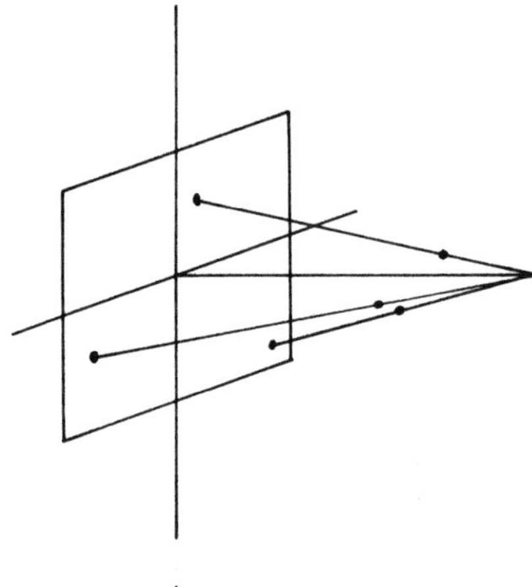

The choice of the distance d determines the angle of view for the picture and should be based on the size of your picture frame, as measured in the (x, y) coordinate system. Suppose that the width of your picture frame is w units; then the following rules for d are based on the standards used in photography:

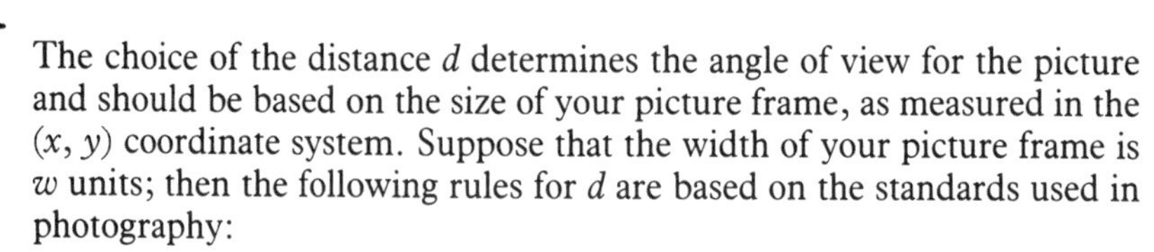

View angle	Wide	Standard	Telephoto
d	$w/2$ to $2w$	$2w$	$2w$ to $10w$

For most purposes, the object to be projected should lie no nearer to the viewer than the picture surface and usually somewhat further away. That is to say, the z coordinates of the object should all be a negative value. This precaution is designed to avoid over-distorted effects of foreshortening. However, you can experiment to get an idea of what is possible.

In order to view an object from various distances, from various angles and in various states of magnification, you will need to be able to translate, rotate and scale its coordinate values. These operations are the subject of the next section.

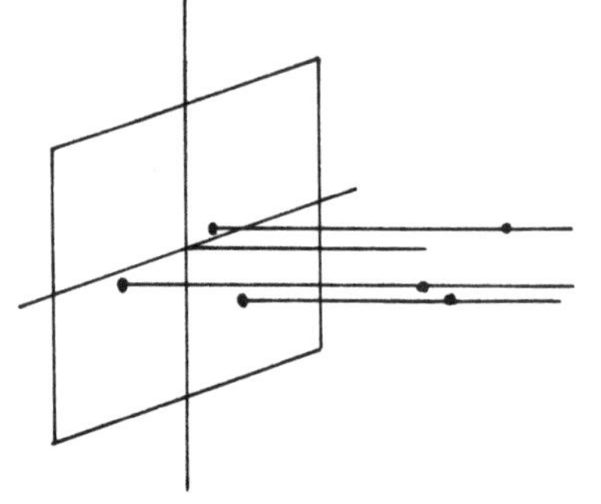

Figure 5.2A

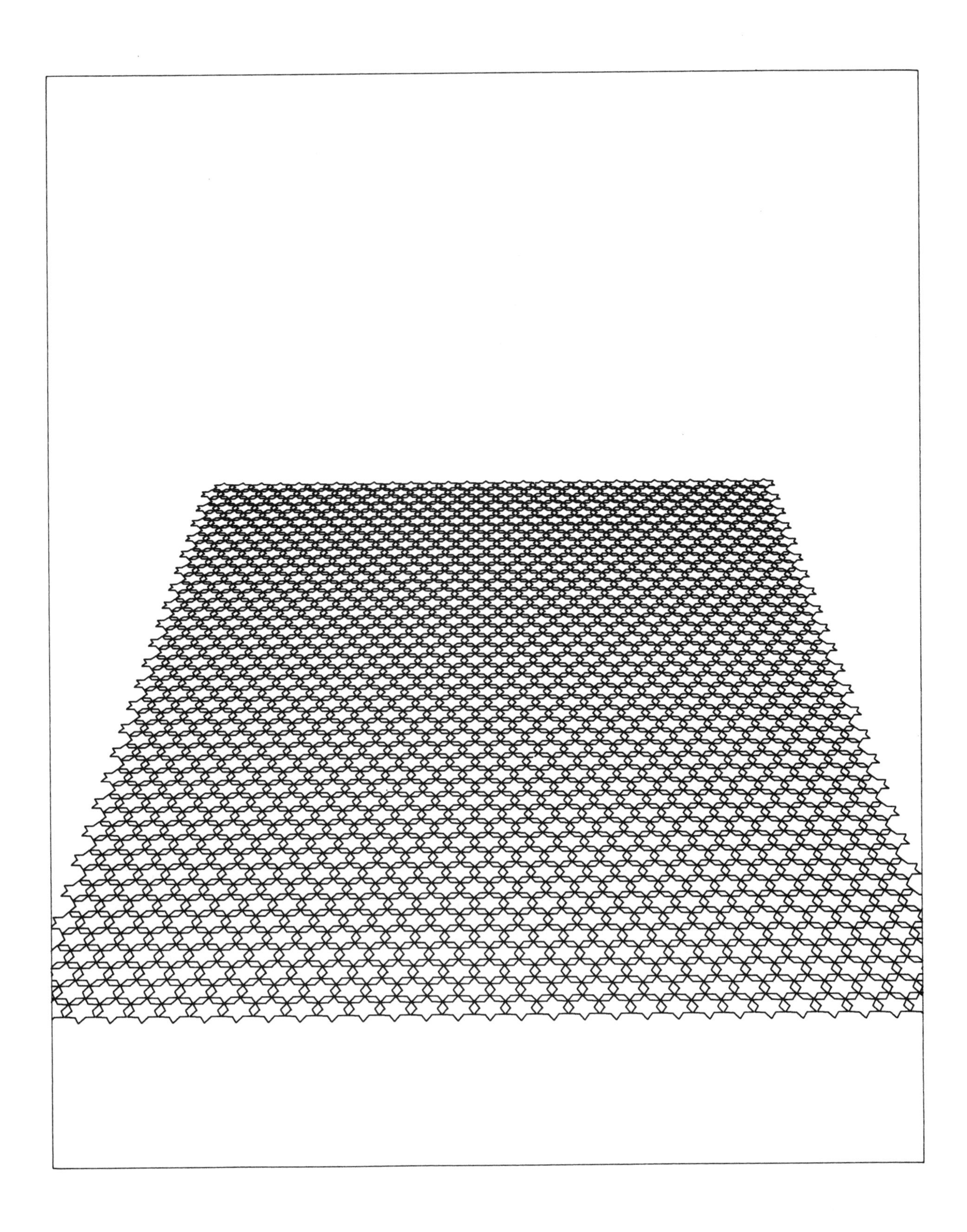

63 Plane in perspective

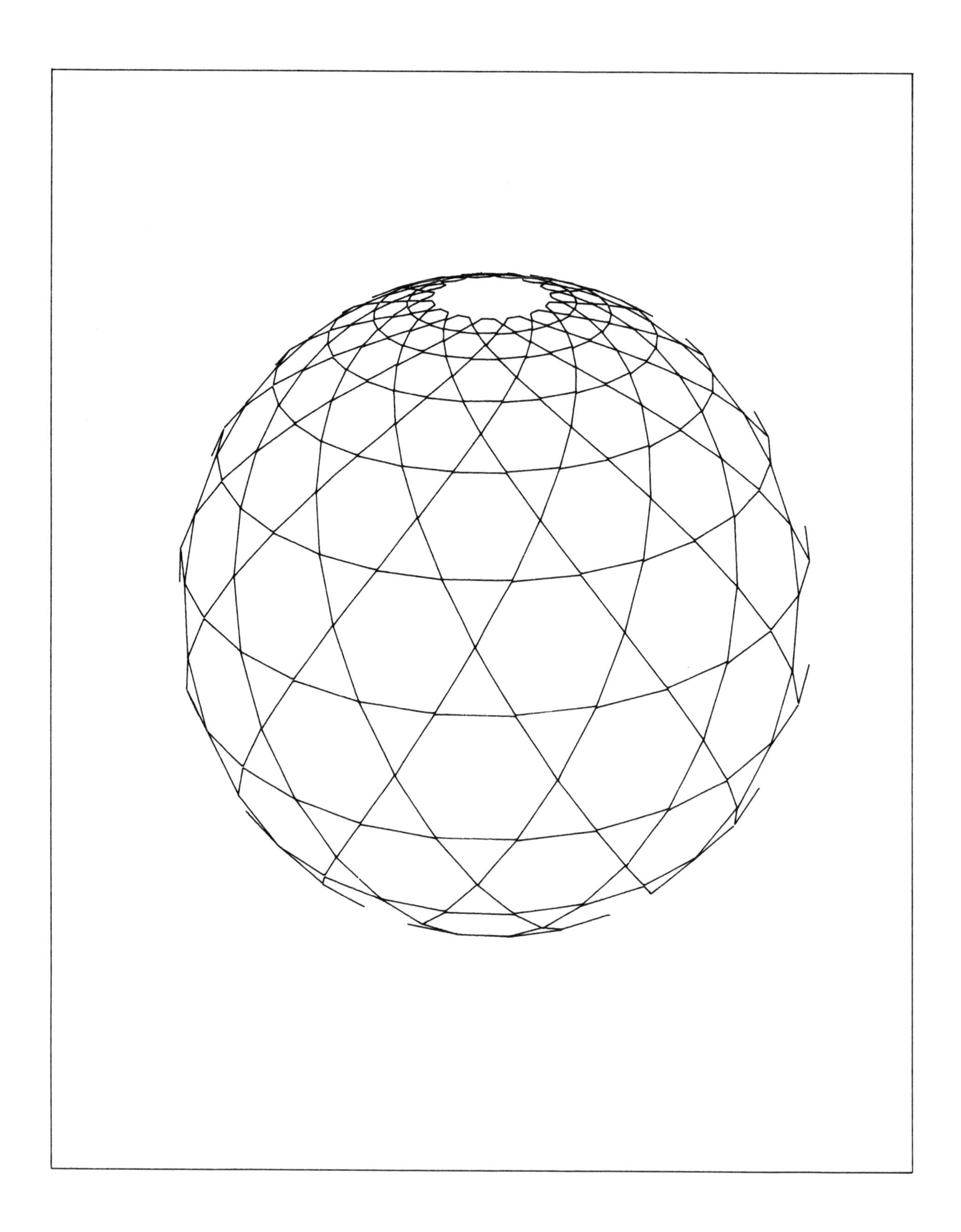

64 Sphere in perspective

5.3 Ten elementary transformations

Although BASIC and traditional algebra are so much alike that the ability to use one leads quickly to the ability to use the other, there is a large difference of meaning in the = sign. In algebra this means 'equals', but in BASIC it sometimes means 'becomes'. *Geometric transformations* are performed on an object by systematically changing the values of (X,Y) or (X,Y,Z) for each point:

```
FOR N=1 TO ...
X= ...
Y= ...
Z= ...
NEXT
```

The purpose of this section is to give BASIC formulae for some of the simplest transformations, beginning with the movements of an object and going on to the problem of mapping a plane onto a sphere.

TRANSFORMATION 1: THE IDENTITY X=X:Y=Y (:Z=Z)

A transformation which alters nothing may seem like a contradiction in terms and certainly quite useless, but not so. It is as important to transformation geometry as 0 is to arithmetic. Geometers are particularly interested in one-to-one transformations, so that every transformation has a corresponding *inverse* transformation which restores all points to their original location. Geometers are also interested in the *products* of transformation; one transformation followed by another transformation is also a transformation. The product of any transformation and its inverse is the identity.

With the exception of the last two (transformations 9 and 10), which involve cutting and pasting, all the following transformations are of a one-to-one nature. The next three (transformations 2–4) are the simple object motions, translation, rotation and reflection. They are collectively known as the *isometries*, or else *congruent* transformations, because they do not alter lengths in the object.

TRANSFORMATION 2: TRANSLATION (DRAWING **65**)

Translation is represented by

X=X+A : Y=Y+B (:Z=Z+C)

It has the following characteristics: it changes location; it preserves size and orientation.

TRANSFORMATION 3: ROTATION (DRAWING **66**)

Rotation by an angle of *A* about the coordinate origin (anticlockwise) is represented by

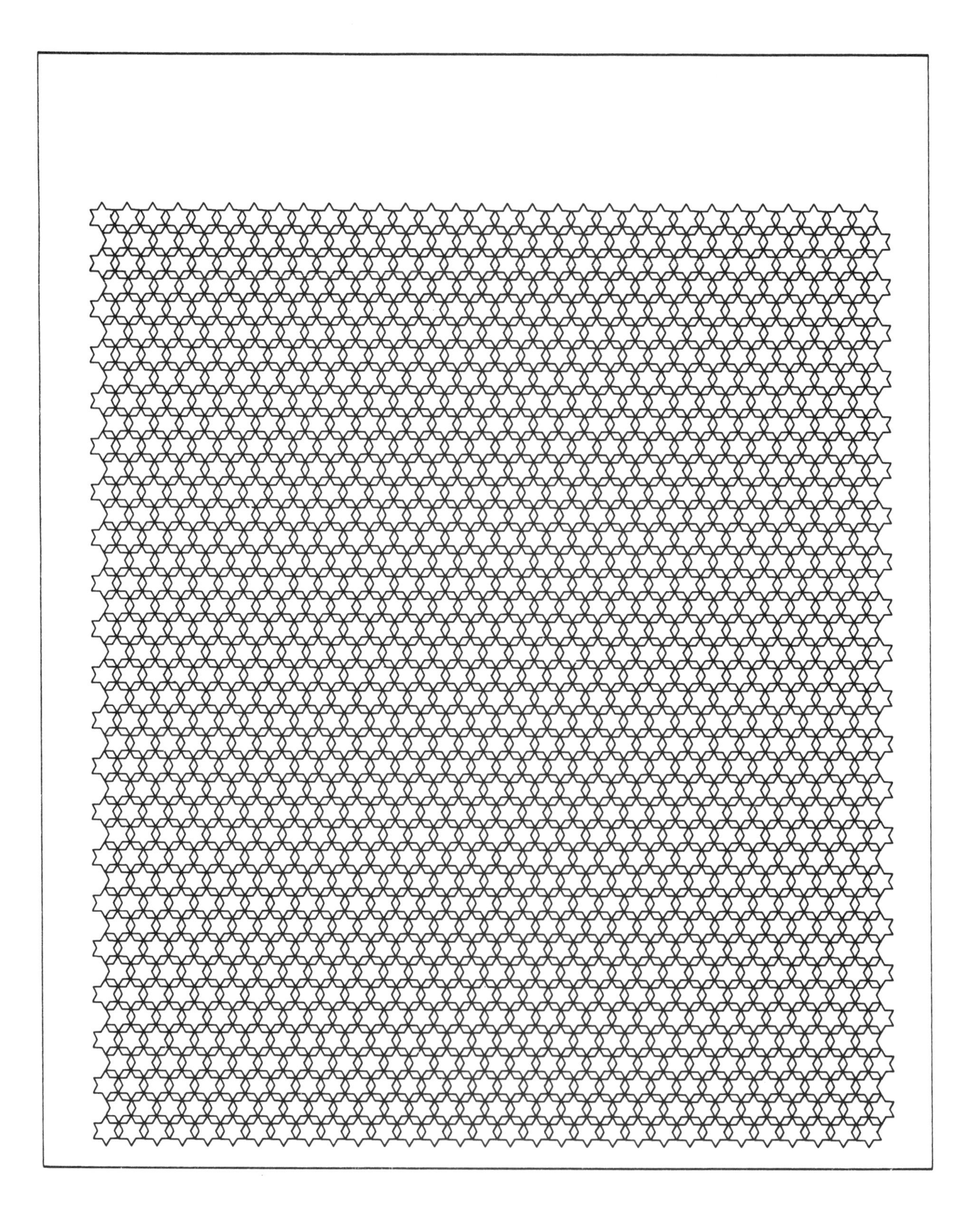

65 Translation

66 Rotation and translation

X2 = X1*COSA − Y1*SINA
Y2 = X1*SINA + Y1*COSA

If you are working in three-dimensional coordinates, the above formula is a rotation about the z axis. Rotations about the x or y axes are obtained by interchanging Z with X or Y respectively. Rotation about any other line through the coordinate origin can be achieved by a combination of rotations about the coordinate axes.

To rotate about any other point in the xy plane, say (XC,YC),

translate (X = X − XC : Y = Y − YC)

rotate about coordinate origin

translate (X = X + XC : Y = Y + YC)

This has the effect of moving the origin of the coordinate system to the centre of rotation before the rotation and back again afterwards.

TRANSFORMATION 4: REFLECTION (DRAWING 67)

Reflection in the y axis (or zy plane) is represented by

X=−X

The complexities of reflecting in other lines (or planes, if you are working in three dimensions) are not pursued in this book, but the details are left as an exercise.

Reflection in a line consists of moving all points to their mirror image in the mirror line. The line joining a point to its image is perpendicular to and bisected by the mirror line, and similarly for reflection in a plane. The farmhouse seen upside down in the mill pond lies underground directly below the real farmhouse. The plane of the mill pond surface cuts the lines joining each point and its image exactly in two and at right angles.

Reflection interchanges left-handedness with right-handedness, and clockwise with anticlockwise. It follows that the product of two successive reflections cannot be a reflection. The details of how two reflections can make the identity, a translation or a rotation are left for you to puzzle out. While you are looking into that, you should go on to consider all other possible products of isometries. When, for example, do two rotations make a translation?

Finally, there is reflection in a point (which is sometimes known misleadingly as *central inversion*) in which the line joining each point to its image is bisected by the mirror point: X = −X : Y = −Y (:Z = −Z).

As well as providing us with basic object manipulations, the three isometries play the key role in analysing symmetry patterns. A butterfly, for example, *coincides with its own mirror image* when reflected in its mid-line (plane), a property known as mirror symmetry. Rotational symmetry and translational symmetry occur when an object coincides with its own image after rotation and translation respectively. An object can have several symmetries together, but there are strict rules governing the possible combinations. There are, for example, only 17 different patterns of symmetry (combining reflection, translation and rotation) in the plane which repeat in two directions. Artists at one time or another have discovered all of them and used them as the basis of decorative designs. Look at wallpapers, mosaics, textiles and carpets for different examples.

67 Chessboard reflected in a sphere

TRANSFORMATION 5: SCALING

Scaling by a factor of K about the coordinate origin as centre is represented by

X=K*X : Y=K*Y (:Z=K*Z)

For $1 < K < \infty$ the effect is magnification, while $0 < K < 1$ gives diminution.

To scale about another point as centre, say (XC,YC),

translate (X=X−XC : Y=Y−YC)
scale
translate (X=X+XC : Y=Y+YC)

Scaling alters size but not shape.

TRANSFORMATION 6: AFFINITY

Affinity is represented by

X=A1*X+B1*Y
Y=A2*X+B2*Y

This class of transformations includes shearing and stretching. We do not pursue it here but merely note that affinities preserve straight lines, parallels and proportions in any given line and that affinities alter angles, shape, proportions not in a line, and sizes.

TRANSFORMATION 7: PROJECTION

Projection was dealt with in Section 5.2. Projection preserves straight lines, but not parallels, nor proportions in any line.

TRANSFORMATION 8: INVERSION (DRAWINGS **68–74**)

Inversion in a circle of radius R centred at the coordinate origin is represented by

X=X*R↑2/SQR(X↑2+Y↑2)
Y=Y*R↑2/SQR(X↑2+Y↑2)

This has the effect of turning the xy plane inside out around the circle of inversion as described in Section 1.6. When working in three dimensions, we can perform a space inversion, which has the effect of turning space inside out about a fixed sphere of inversion, by

X=I*X : Y=I*Y : Z=I*Z

where

I=R↑2/SQR(X↑2+Y↑2+Z↑2)

To invert about a circle centred elsewhere, say at (XC,YC,ZC),

translate (X=X−XC : Y=Y−YC : Z=Z−ZC)

invert about a circle centred at the coordinate origin

translate (X=X+XC : Y=Y+YC : Z=Z+ZC)

Inversion turns lines into circles and planes into spheres. To map the

68 Inversion

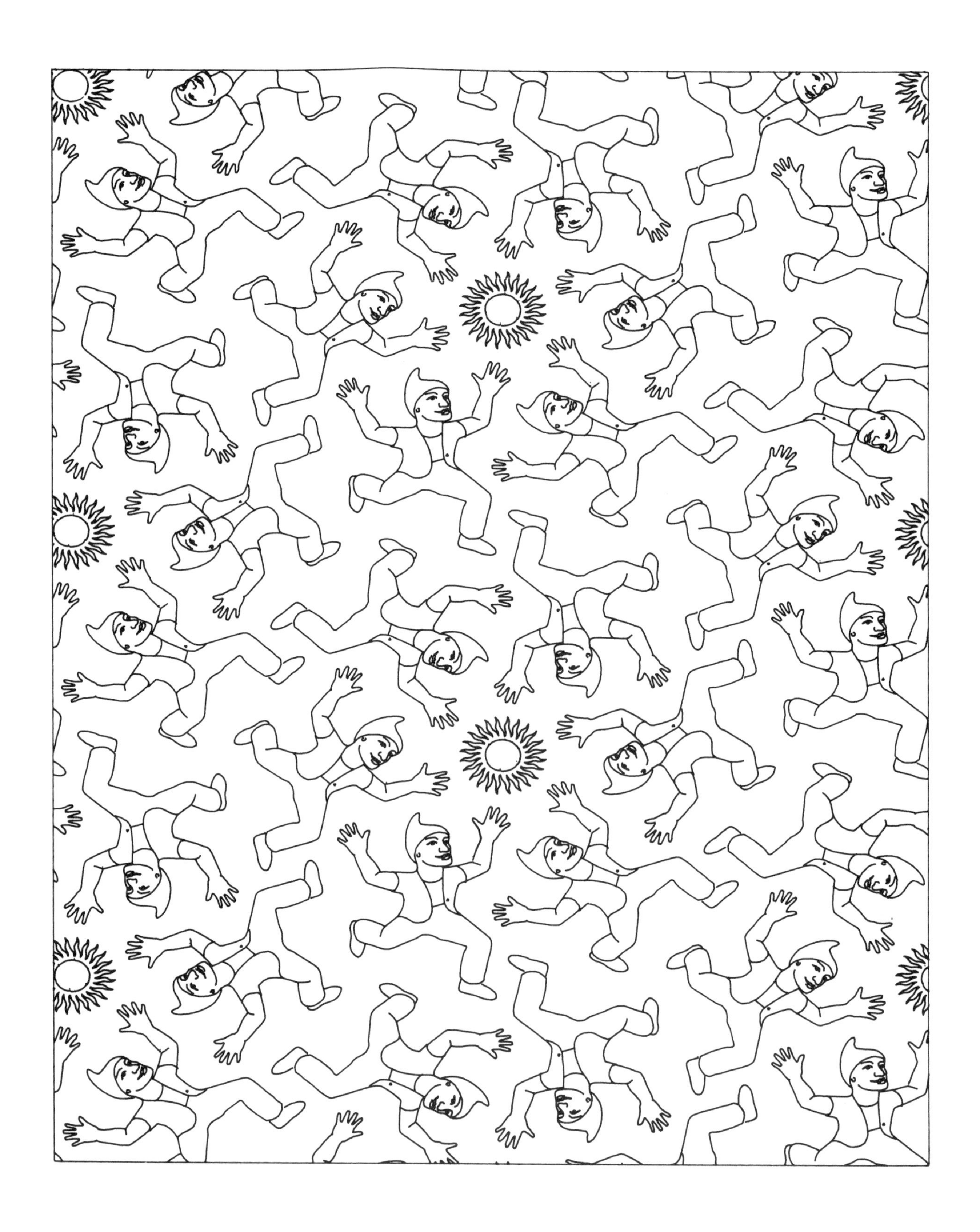

69 Men on a plane

70 Men on a sphere

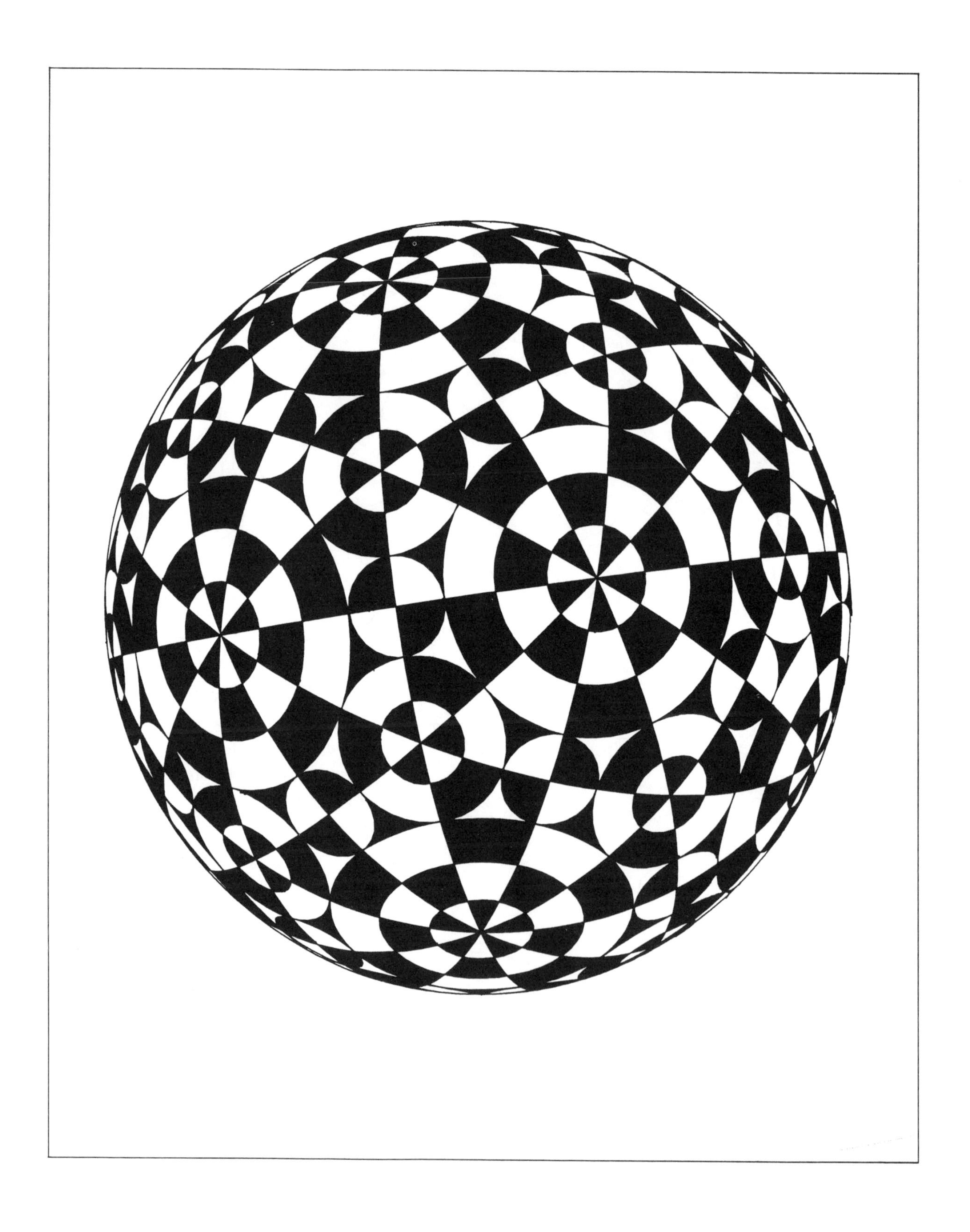

71 Spherical symmetry

72 Inversion of circles between parallel lines

73 Inversion of a chessboard about its centre

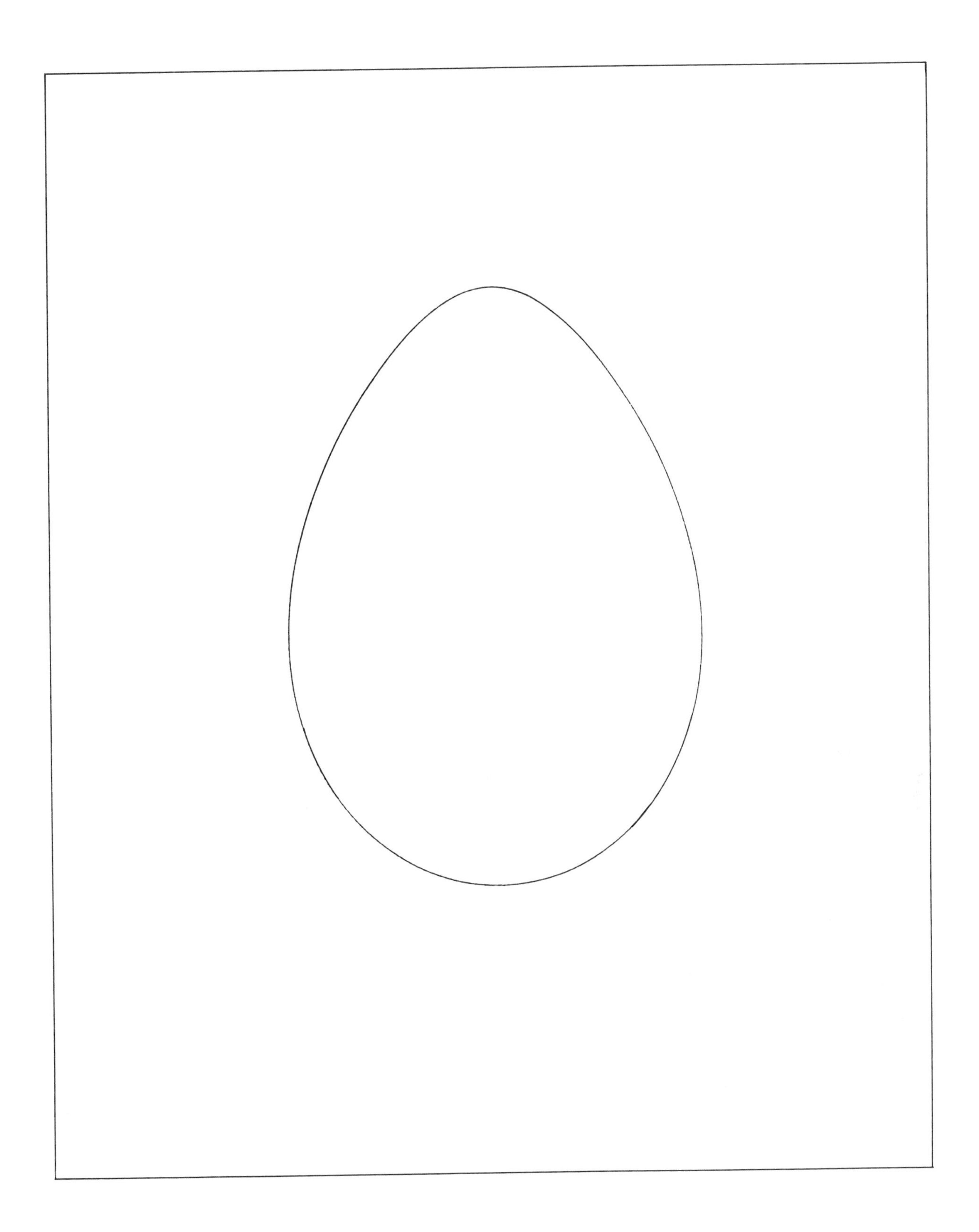

74 Ellipse inverted about external point on its major axis

Figure 5.3A

xy plane into a sphere whose poles are at (0, 0, 0) and (0, 0, D), perform the inversion in the sphere of radius D centred at (0, 0, D).

Figure 5.3A shows an important property of inversion, the fact that it does not alter angles. See the points of the stars. Transformations with this property are called *conformal.*

Inversion is also, like reflection, a sense-reversing transformation, interchanging right- and left-handedness. For further discussion on inversion, see Section 1.6.

TRANSFORMATION 9: THE ANTIMERCATOR (DRAWINGS 75–80)

The antiMercator is given by

```
A=K*X
R=EXP(K*Y)
```

This is a conformal mapping of part of the plane, the endless strip bounded by $x = 0$ and $x = 2\pi/K$, onto the whole plane. Horizontal lines (y = constant) become circles concentric with the coordinate origin. Vertical lines become radial to the coordinate origin. Slanting lines become logarithmic spirals about the coordinate origin. As the image under this transformation is given in polar coordinate form, you will need to convert to Cartesian by

```
X=R*COSA : Y=R*SINA
```

75 Row of ellipses after an antiMercator

76 Rows of identical stars after an antiMercator

77 Inversion of Drawing **76**

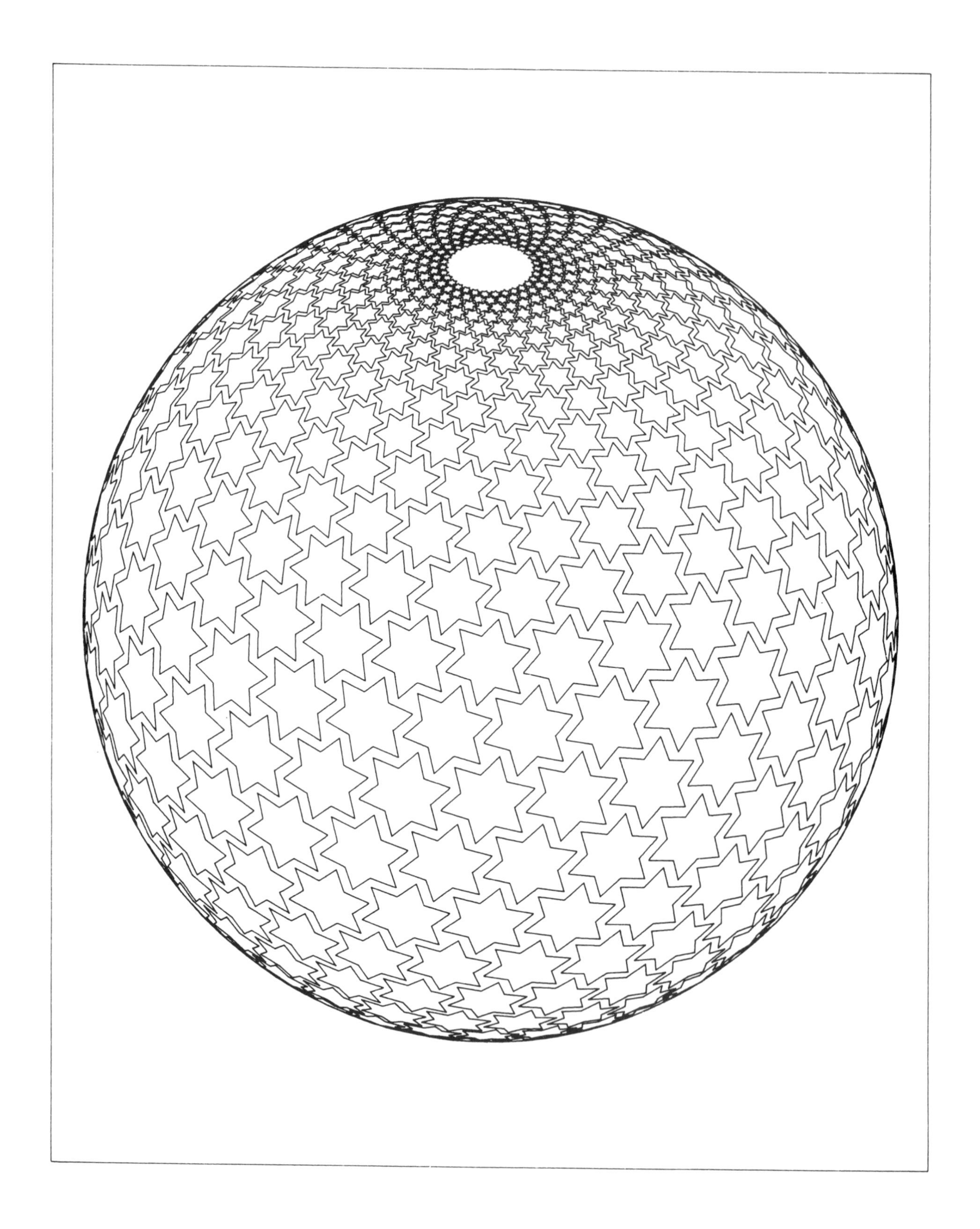

78 Rotated view of Drawing **77**

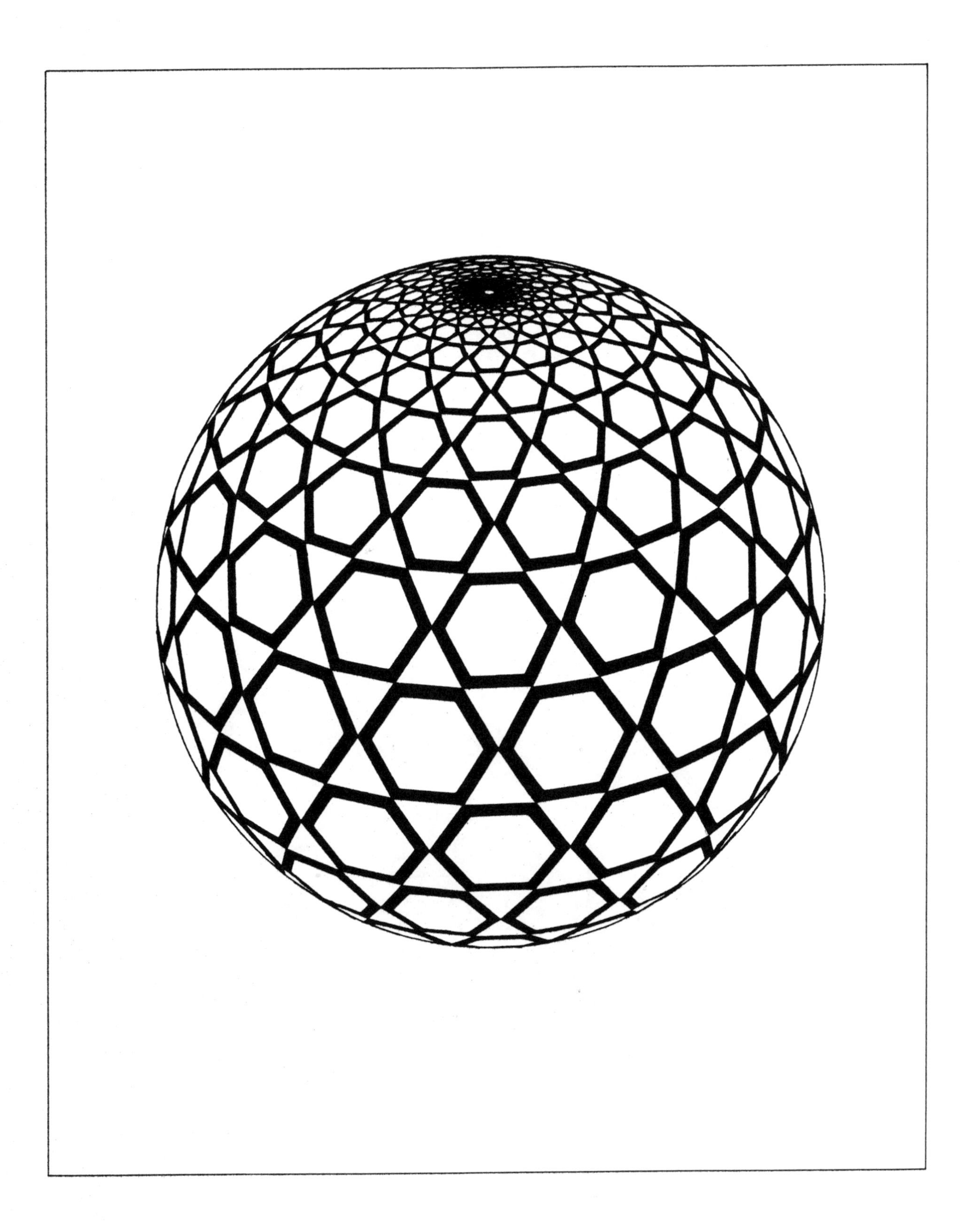

79 Sphere

80 Drawing **79** inverted into a plane

I have here used the antiMercator to turn a row of touching ellipses into a ring of touching eggs, and to turn a pattern of stars laid out in parallel rows into one laid out in concentric rings. This can then be inverted into the sphere as illustrated and described above.

The reverse of this transformation is the *Mercator*:

X=A/K : Y=LN(B)/K

where $K=2\pi$/width of strip. Mercator's map of the World published in 1569 and still widely used today is obtained by applying a Mercator to a *stereographic projection* (inversion of the planet about a sphere centred at the north (or south) pole and of radius equal to the Earth's diameter). It greatly exaggerates polar regions of the globe, such as Greenland, by comparison with some preferred latitude, where all the proportions are faithfully preserved.

TRANSFORMATION 10: CYLINDERWRAP

Cylinderwrap is represented by

X=SIN(K*X)/K
Z=COS(K*X)/K
(Y=Y)

This rolls up the strip of the *xy* plane lying between $x = 0$ and $x = W$, where $W = 2\pi/K$, into a cylinder around the *y* axis.

Exercise 17

1 State the condition under which two reflections combine to make the following.

a The identity. **b** A translation. **c** A rotation.
d A reflection. **e** A half-turn.
Draw diagrams to illustrate your answers.

2 State the condition under which two rotations make the following.

a The identity. **b** A translation. **c** A rotation.
d A reflection.
Draw diagrams.

3 What transformation is equivalent to when two inversions have the same centre but different radii?

4 Under the action of an anti-Mercator where is the image of the line $y = -\infty$?

5 Are the following preserved under an anti-Mercator?

a Angles. **b** Circles.

6 What is the image of a diagonal line such as $y = x$ under the following?

a An anti-Mercator. **b** A cylinderwrap.

7 Find the formulae for a reflection in the following lines.

a $y = x$. **b** $y = x \tan A$.

8 Under the action of the translation X=X+A : Y=Y+B : Z=Z+C, how far is each point moved?

9 Construct a pattern of icosahedral–dodecahedral symmetry on a sphere using the following information. Let there be 12 poles, each with fivefold rotational symmetry, in six antipodal pairs; if the first pair is aligned with the z axis, the other five are found by a rotation of 1.1072 radians about the y axis, followed by a rotation of $2\pi n/5$ radians about the z axis (n = 0, 1, 2, 3 and 4). .

Reading

Coxeter, 1961; Hilbert and Cohn-Vossen, 1952; Lockwood, 1961; Markushevich, 1962; Resnikoff and Wells, 1984; Stevens, 1981; Weyl, 1982.

5.4 The fractal universe

Big whorls have little whorls,
Which feed on their velocity,
And little whorls have lesser whorls,
And so on to viscosity.

With these words based upon Swift's verse on fleas, L.F. Richardson describes the shape of *turbulence* in, for example, water. A complexity such as this, resulting from the simultaneous occurrence of events on many different scales of size, is called a *fractal.* Some other examples follow.

a Moons orbit planets, which orbit stars, which orbit the centres of galaxies, which orbit the centres of galaxy clusters, ...
b Each branch of a tree is itself like a miniature tree, with its own branches, which in turn have their own branches, and so on until you come to the leaves.
c A river is like a tree, fed by tributaries, which are in turn fed by smaller streams, until you reach the smallest dribbling brooks.
d A coastline is made of promontories and inlets, each of which is further shaped by smaller promontories and inlets, and so on down to the pebbles.
e There are some islands which have lakes on which there are islands which have lakes.
f The waves on the sea are themselves wavy with smaller waves, and so on down to the smallest ripples.
g The shape of clouds in the sky, which show remarkable similarity when viewed at scales ranging from whole-earth satellite photographs down to fragments of swirling mist seen on mountain walks, reveals the turbulence of the atmosphere.
h The outline of a mountain is an irregular bump which, on closer inspection, turns out to be composed itself of smaller bumps. The thrusting, twisting and faulting of rock in the Earth's crust happens simultaneously at many levels of scale.
i The craters on the Moon are themselves pitted with smaller craters, which are also pitted. These roughly circular impact scars range in diameter from over 1000 km to about 1 m.

Some of these patterns are highly random, such as the locations and sizes of lunar craters, while others are more orderly. Some, such as a tree, whose shape meets the conflicting demands of light, air, wind and gravity in optimal fashion, are very clearly ordered. Each phenomenon is strictly limited between an upper and a lower scale of size, but sometimes, as in the example of orbiting bodies, the range of scales included is unimaginably large. A striking similarity of form or event at different levels of scale is also a notable feature of the above patterns.

ARTISTIC FRACTALS

Pictures and patterns often rely for their pleasing effect upon organisation of parts in a fractal manner, larger shapes subdividing into smaller shapes which further subdivide. Architecture provides good examples of this hierarchical principle. In our appreciation of texture and composition, we look for features to be present at every visible scale.

GEOMETRIC FRACTALS (DRAWINGS **81–84**)

It was the mathematician Benoit Mandelbrot who coined the name *fractal* to describe the many mathematical curves and surfaces which behave like the above-listed patterns. The important difference between a natural fractal and a mathematical fractal is that the latter is conceived as continuing to show detail at every level of magnification. A *fractal curve* is one which has infinitely many bends between any of its two points. A *fractal surface* is one which has infinitely many hills and dips in any locality. A *fractal dust* is one in which every particle turns out to be a cloud of particles... .

Of course, in attempting to draw any of these forms, we shall only run to a finite number of discernible features. The following drawings are geometric fractals, ranging from the highly abstract to the quite naturalistic.

81 Fractal tetrahedron (becoming a dust)

82 Rough surface

83 Nested circles

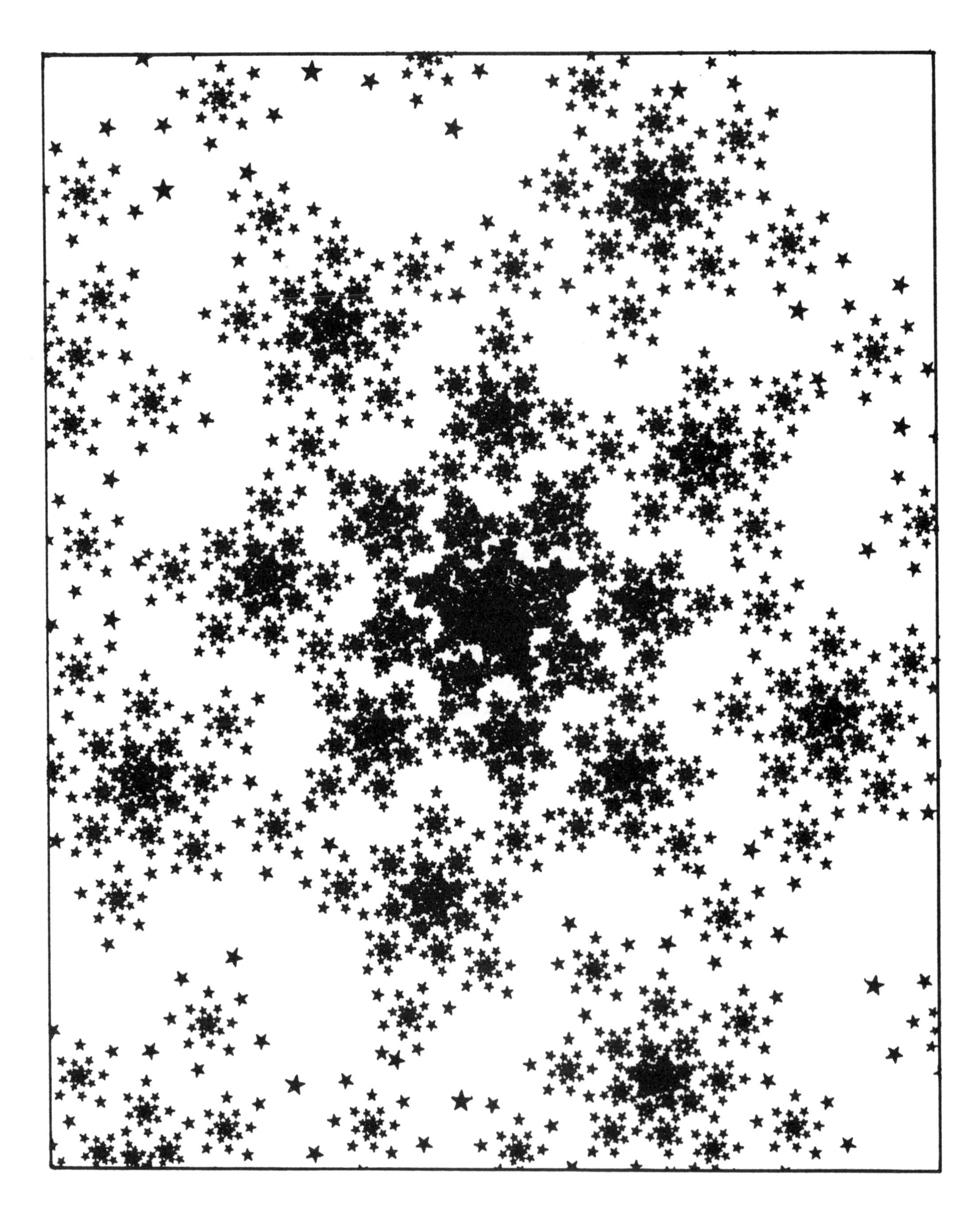

84 Fractal phyllotaxis: cauliflower

Exercise 18

These drawings happen also to pose a range of computing problems for the reader to solve. The notes provided describe the pattern in each case and offer programming clues. Program outlines can be found in the Answers section for those who have tried without success for more than a week!

Draw the following

1 The blancmange curve.

2 The Koch curve.

3 Dirichlet's function.

4 Pentasnow.

5 Condensation.

6 Lunar craters.

7 Water waves.

THE BLANCMANGE CURVE (DRAWING **85**)

Figure 5.4A shows how this curve is created stage by stage. You can imagine it to be a perfect piece of elastic stretched between two fixed points, the end points of the curve. Stage 1 consists of raising the centre of this single line by an amount proportional to the line width, and fixing it there. This creates two line segments. Stage 2 consists of raising the centres of each of these two line segments by half the amount in stage 1 to create four line segments, and so on.

David Tall uses this curve to illustrate the mathematical idea of a curve which has no definite slope (*derivative*) at any of its points. For, if the above-illustrated process of bisection is continued forever, then you end up with corners everywhere!

Fortunately, we shall only need to carry the algorithm to about a dozen stages before the further effects become no longer visible in a drawing.

Drawing **85** shows the blancmange curve at the tenth stage of bisection and has $2^{10} = 1024$ points joined by straight lines.

To program this curve, create an array for the heights of 1024 points, and compute them in the order Y(512), Y(256), Y(768), Y(128), Y(384), Y(640) and so on.

THE KOCH CURVE (DRAWINGS **86–90**)

In his long story about how nature is fractally shaped, Mandelbrot starts with the curve first proposed by Koch in 1904 as an idealised coastline with triangular promontories and bays. Suppose that every straight segment turns out to be two bays and a promontory on closer view, as shown in Figure 5.4B.

What Koch had noticed was that this curve becomes infinitely long. Each stage of the construction increases the total length of the curve by the same proportion. It also illustrates the idea of *self-similarity*; any part of the curve is a scaled-down version of the whole curve.

If you have succeeded in programming the blancmange curve, the Koch curve and its numerous variants present only one further geometric problem to solve. Given a line segment whose end points are (x_1, y_1) and

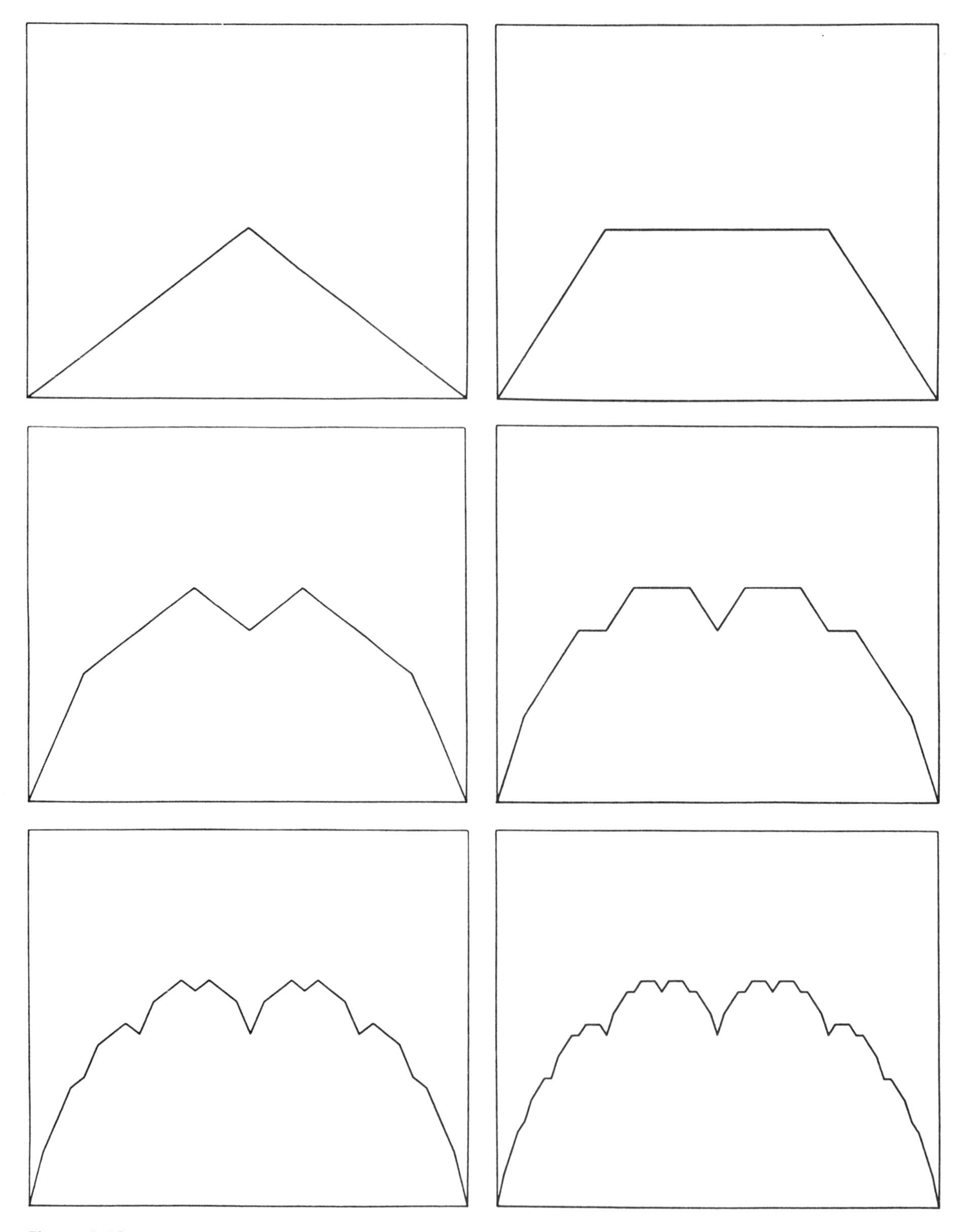

Figure 5.4A

(x_2, y_2), find the coordinates (x_3, y_3) of its mid-point after being displaced perpendicularly to the segment by an amount of K times the length of the segment. The answer is given in the Answers section. The mid-point can be displaced left or right of the line, and you should try to vary the direction from stage to stage and from segment to segment, to get the various curves shown here. The effect of varying the numerical value of K is to alter the corner angles, which corresponds to a range of wiggliness which runs from smooth to impossibly rough.

Figure 5.4B

THE DIRICHLET FUNCTION (DRAWING 91)

Whole numbers and ratios of whole numbers, such as 3/7, are called *rational* whereas numbers such as $\sqrt{2}$ or π, which cannot be written down exactly as ratios of whole numbers, are called *irrational*. According to legend it was Pythagoras who discovered irrationality in numbers when

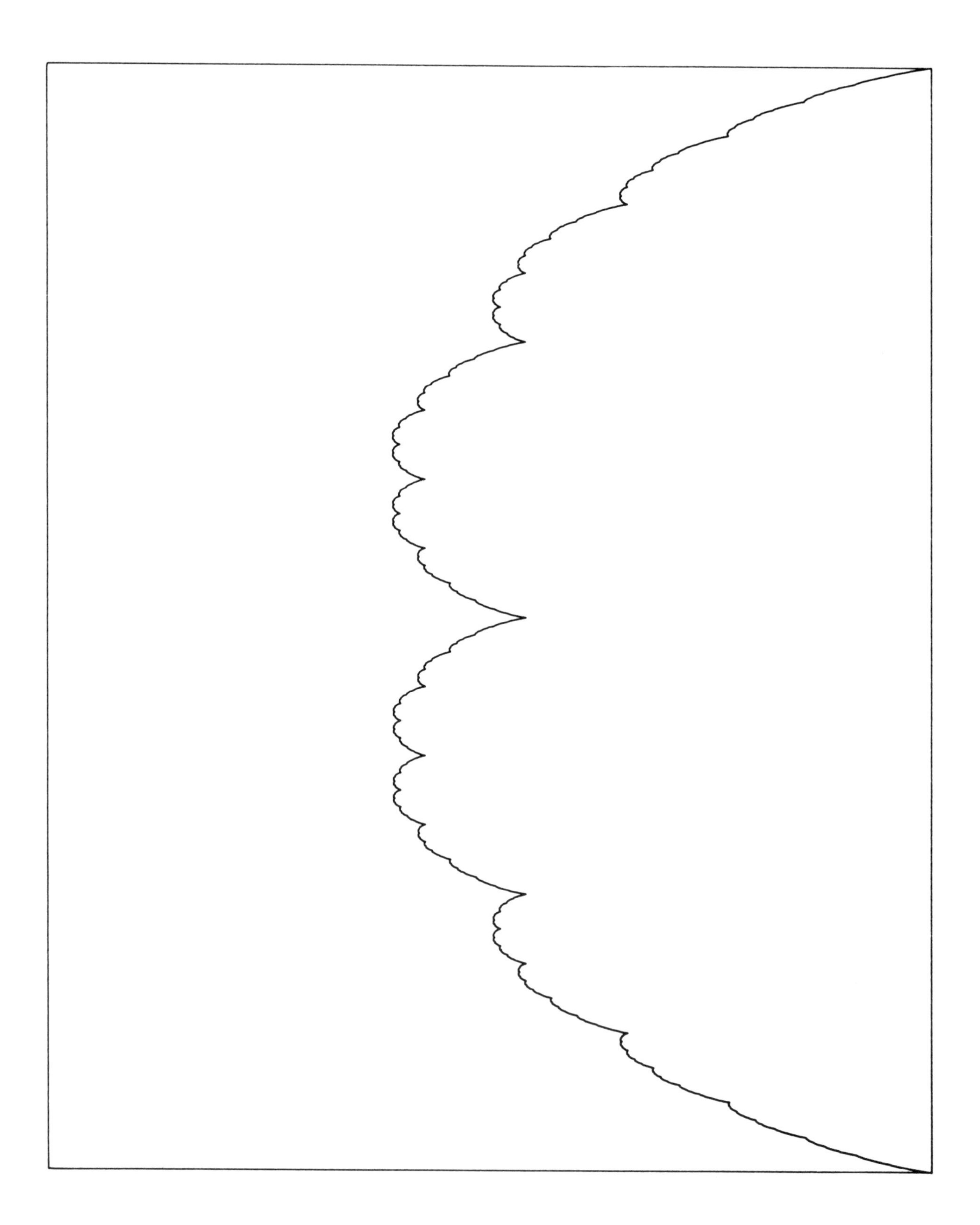

85 Blancmange curve

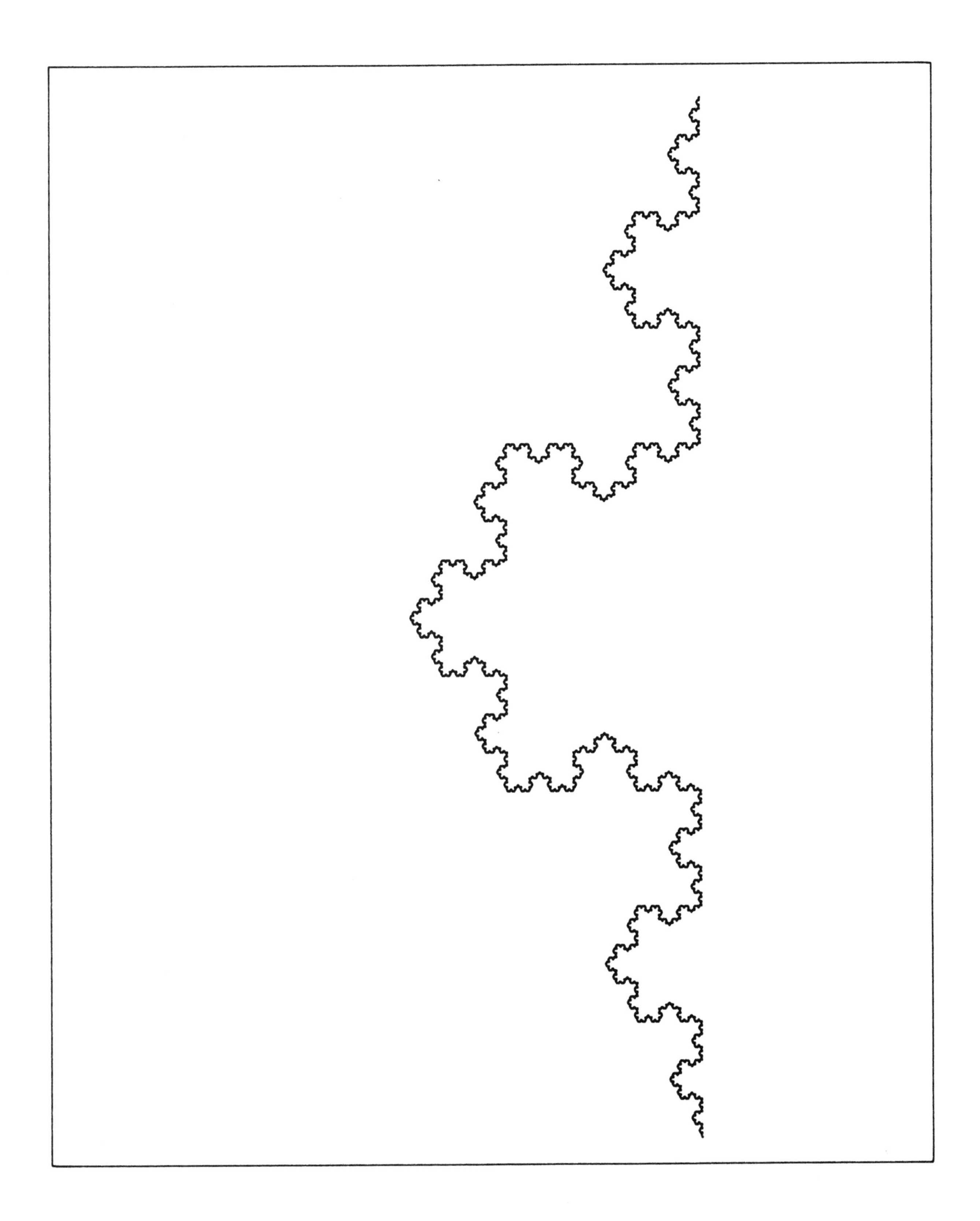

86 Koch curve: displacements alternating each generation

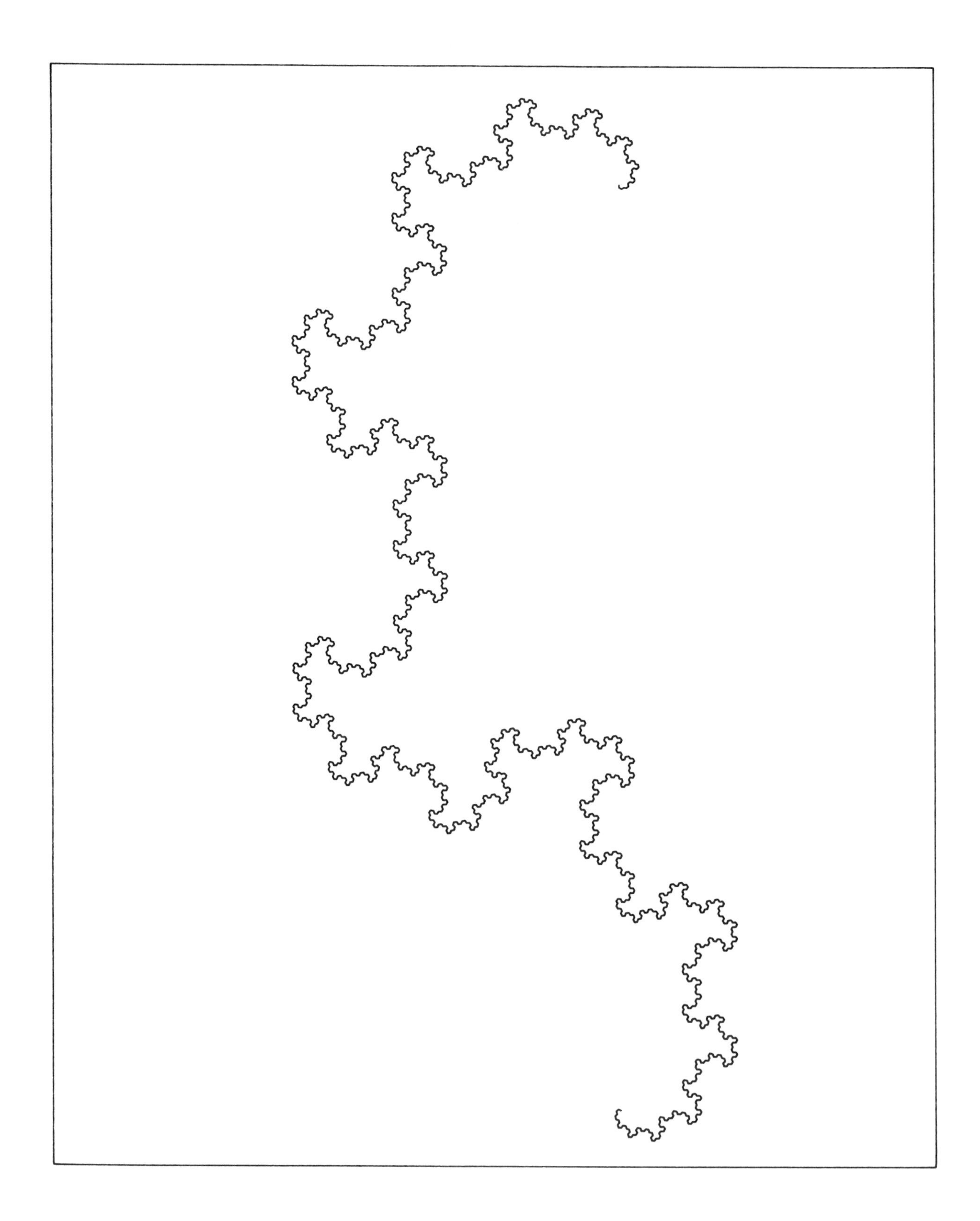

87 Dragon curve: displacements alternating each segment

88 Dragon curve

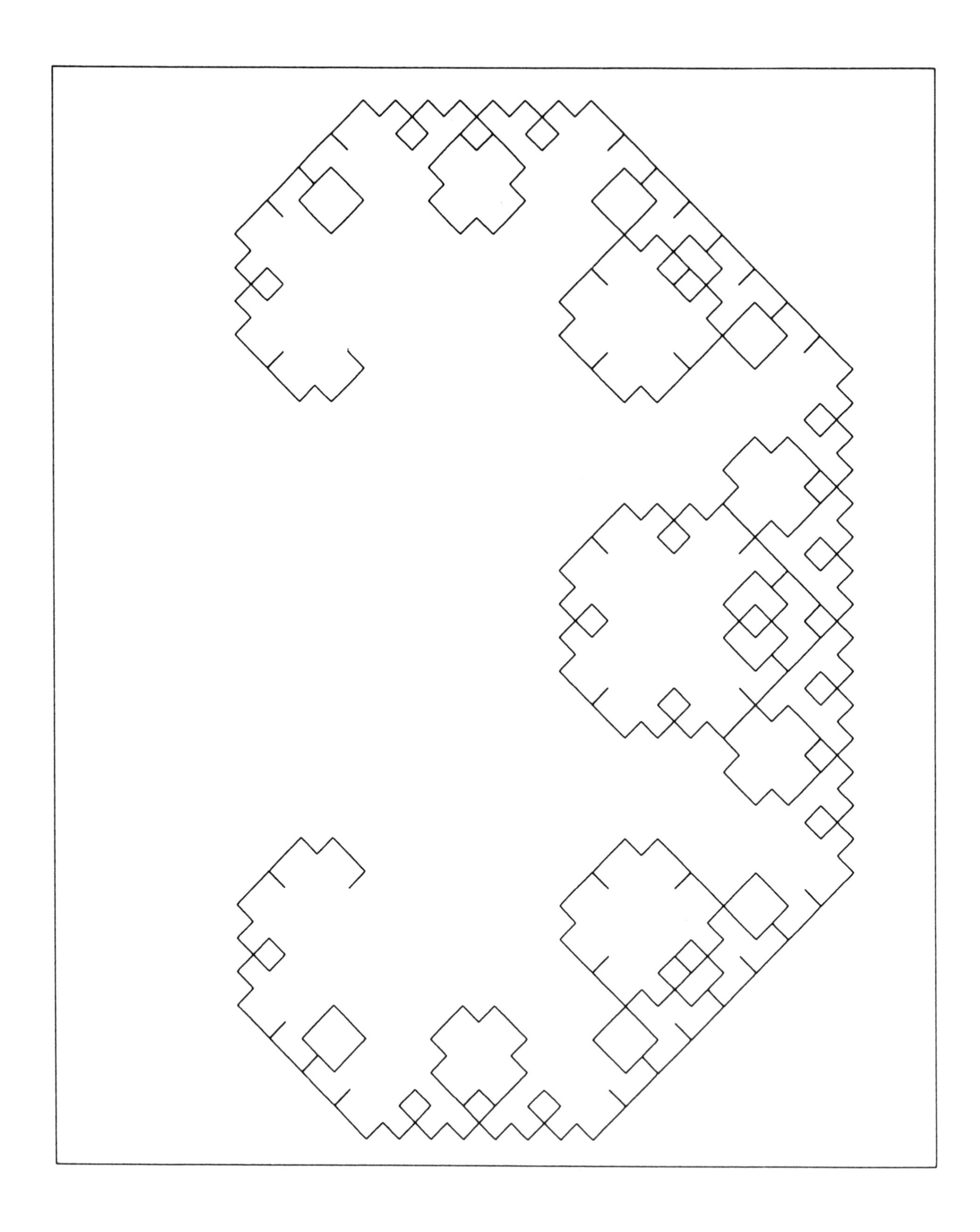

89 C-shaped curve: displacements all in the same direction of $\frac{1}{2}\sqrt{2}$

90 C-shaped curve

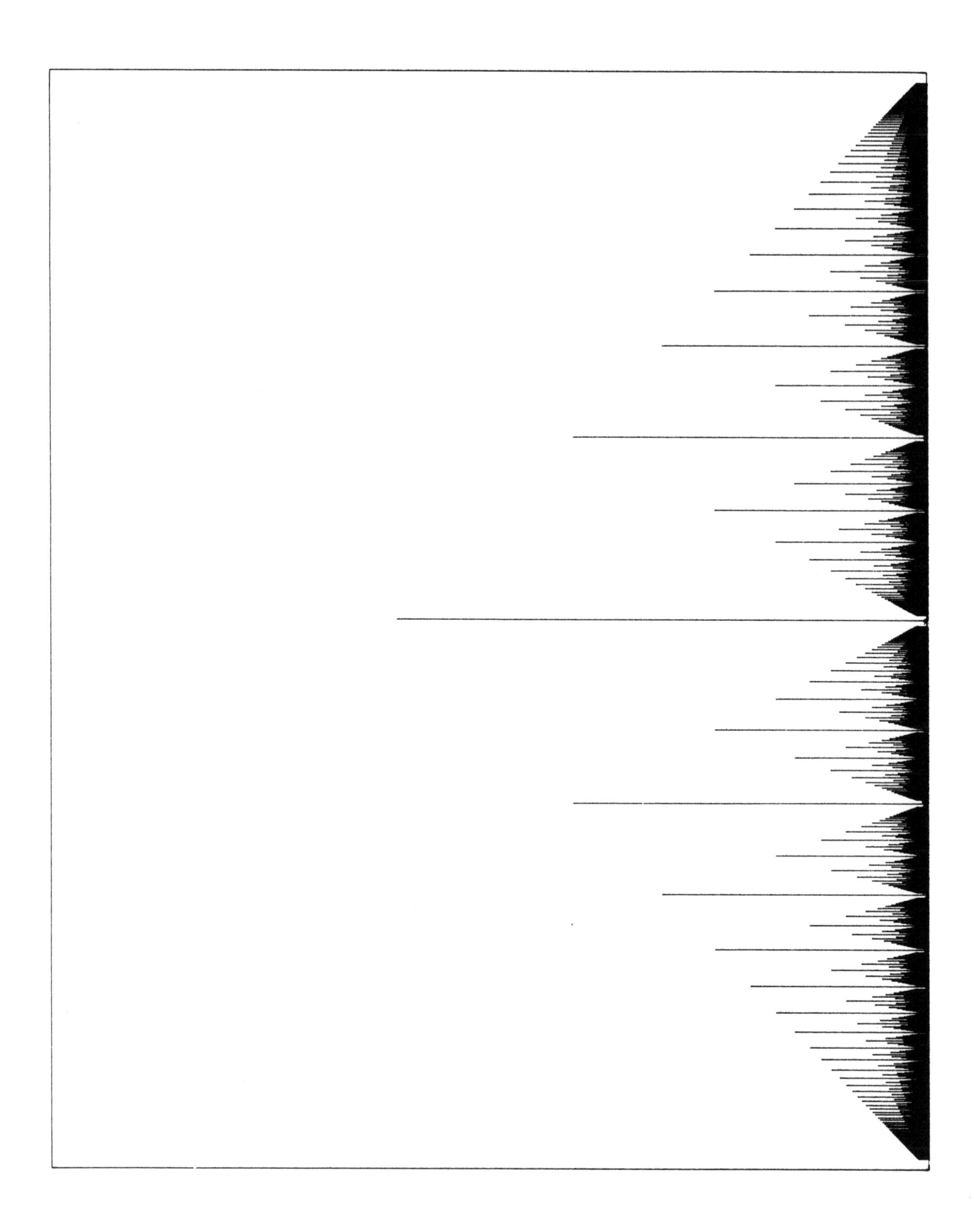

91 Dirichlet's function

he realised that the diagonal of a square could not be commensurable with its side. Suppose that a line marked by 0 at one end and 1 at the other represents all the numbers from 0 to 1. Some of the points will represent rational numbers, such as 2/3, while other points will represent irrational numbers, such as $\sqrt{\frac{1}{2}}$. What is their pattern of arrangement?

First of all, it is easy to show that between any two rationals there is another rational and, similarly, that between any two irrationals there is another irrational. So there is an infinity of both rationals and irrationals between 0 and 1. Indeed the same goes for any interval, no matter how small. To any rational, you can find another rational as close as you wish and, similarly, for irrationals. It is beginning to sound paradoxically crowded. Moreover, there is a famous argument due to Cantor which concludes that the infinity of irrationals is infinitely greater than the infinity of rationals. Finally, between any two rationals there is an irrational, and vice versa.

This state of affairs may be called Dirichlet's paradox, after the nineteenth-century mathematician who pondered it and proposed the following function to illustrate the idea of *discontinuity* with a vengeance:

$$f(x) = 1 \text{ if } x \text{ is rational; otherwise } f(x) = 0.$$

It is quite impossible to illustrate this function properly, because it is supposed to consist of infinitely many separate points. What we can do instead is to *begin* to draw it by drawing in (here indicated as vertical lines) the values for the rationals in the following order: 1/2, 1/3, 2/3, 1/4, 2/4, 3/4, 1/5 and so on, before stopping short of all the lines merging together.

A more interesting version of the Dirichlet function (Figure 5.4C) is given by

$$\begin{aligned} f(x) &= 1/b \text{ when } x \text{ is rational} \\ &= a/b \text{ in fully reduced terms} \\ f(x) &= 0 \text{ when } x \text{ is irrational} \end{aligned}$$

Figure 5.4C

Ian Stewart used this function to illustrate one which is continuous at all irrational points (according to the mathematician's definition of continuity) and discontinuous at all rational points. Perhaps you can see that the pattern which emerges is one of *self-projectivity* (every part of the drawing is a perspective drawing of the whole).

It is an easy and delightful drawing to program, but you should include a subroutine which avoids repeated calls of the same fraction, e.g. 1/3, 2/6, 3/9 and so on.

PENTASNOW (DRAWINGS **92** AND **93**)

Albrecht Dürer was the first to notice that pentagons will pack together in the manner shown here. Six pentagons go together to make one large pentagon, with five holes in the shape of acute golden triangles (Drawing **92**). Six of these pentagons in turn go together to make an even larger and even more holey pentagon, and so on *ad infinitum*.

The pattern is self-similar and, if continued forever, is eventually all holes. The scaling ratio for each pentagon to its six constituents is $\tau^2{:}1$. By a suitable adjustment to the pattern, the holes can be turned into fractal pentagrams each looking like that in Figure 5.4D.

If we could continue this pattern forever, these pentaflakes would eventually join up to cover the entire surface. The artists who use the laws of plane symmetry, especially the Islamic artists, were always trying to find patterns which involved fivefold symmetry. For, unlike the triangle, the square and the hexagon, the pentagon will not fit into a repeating pattern. So here perhaps is a way. The jigsaw of pentaflakes is entirely made up of parts with exact fivefold symmetry. The coverage is asymptotic, however, i.e. it is never completed by a finite number of pieces.

CONDENSATION (DRAWINGS **94–96**)

Imagine that you place a smooth plastic lid on top of a steaming hot cup of coffee and wait a while until the underside of the lid has steamed up with condensation. Lift the lid and look closely. You should see hundreds of tiny droplets, each one separate and circular in outline. The arrangement is not regular, and yet there is a strong tendency for uniform and dense cover. The sizes are not uniform, but they may tend in some random manner towards some preferred size and show clear upper and lower limits to their range.

We shall extract from such observations the following computer problem: to draw randomly scattered circles of random sizes to cover a rectantular area without touching.

Randomness is not chaos; it is variability which is unpredictable in detail, but subject to laws of probability. Different laws of probability generate different patterns of randomness. The most basic, and the one out of which we shall create all the rest, is the *uniform* distribution, which is simulated by the BASIC call of

```
X = RND
```

This will generate a different number on each call whose value lies between 0 and 1, given by a decimal fraction. The probability that x is in any part of this interval would be proportional to the size of the interval. In short,

$$\Pr(x \leqslant n) = n \text{ for } 0 \leqslant n \leqslant 1$$

92 Dürer's pentagons

Figure 5.4D

The program for drawing condensation consists of choosing a location within a rectangle in a uniform random manner, by specifying the coordinates of the centres of circle with

X = RND : Y = RND

At the same time, we must choose a radius for each circle at random,

93 Pentasnow

94 Non-overlapping circles randomly distributed

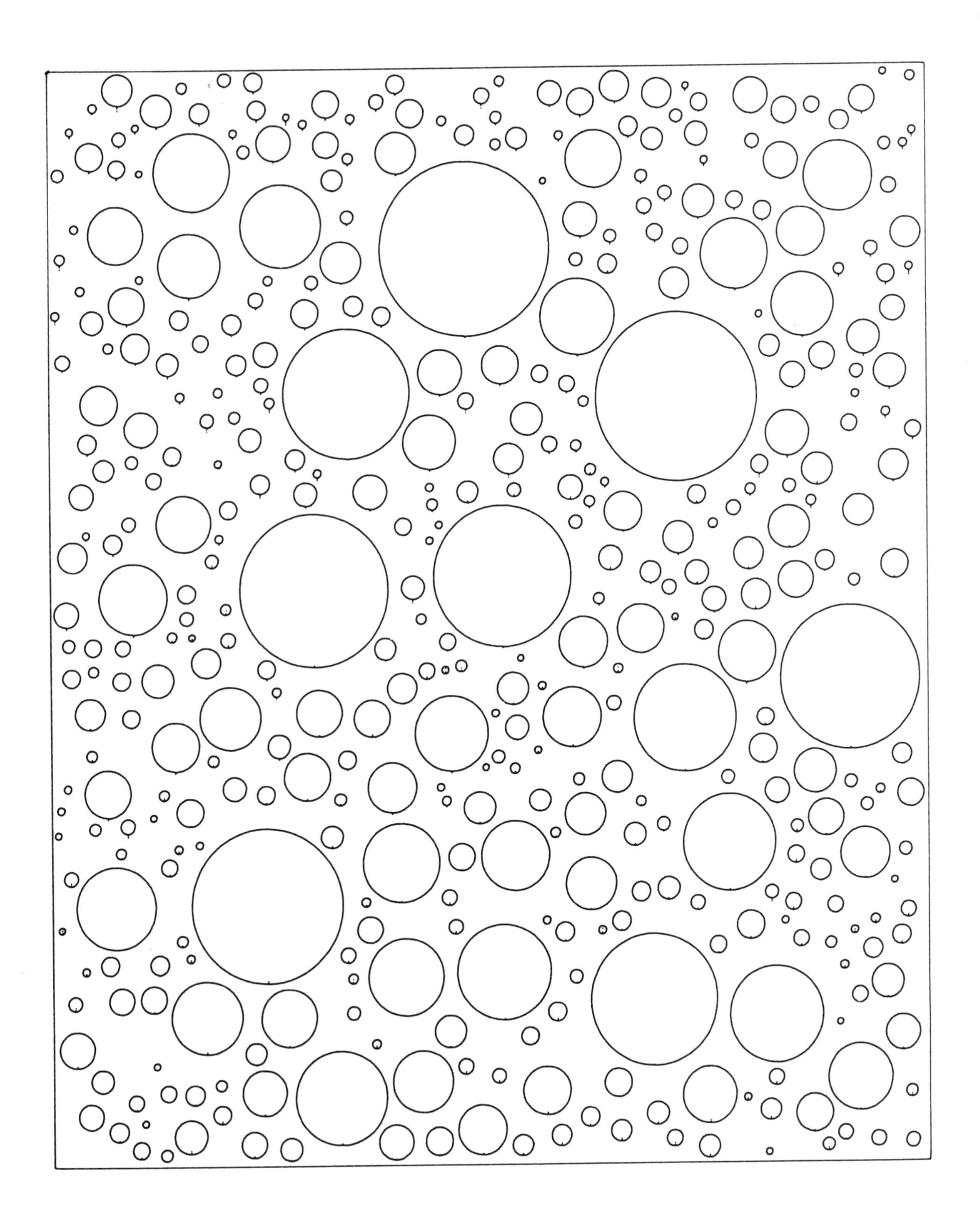

95 Condensation

96 Condensation

but the uniform distribution seems unsuitable. By composing functions of RND, we create other patterns of randomness. For example, R=(RND) ↑ 2 is more likely to produce sizes of radius R nearer to 0 than 1; while R=(RND) ↑ ½ is more likely to produce numbers nearer to 1 than 0. Central tendencies can be created by, for example, R=1+(2*(RND−0.5)) ↑ 3. This gives radii varying between 0 and 2 with a strong tendency to be nearer to 1.

Experiment with forming functions of RND in this way to obtain various patterns of randomness. Further information on this topic is given in the following section on lunar craters.

Having chosen a location for the centre and a size for the radius of a candidate circle in the manner described above, the next job is to make sure that it does not touch or intersect any other circle in the drawing. This is done by generating an array of X(N), Y(N) and R(N) one circle at a time and checking that each circle does not touch or intersect any previously entered into the array. So circles are chosen at random but are recorded in an array only if they pass the clearance test. This test is equivalent to a simple use of Pythagoras' theorem in answering the question: is the distance between the centres of two given circles greater than the sum of their radii?

Drawings **95** and **96** demand some patience as it becomes increasingly difficult to find further circles which pass the clearance test. Readers interested in achieving a greater sense of realism for these drawings are urged to study natural occurrences of condensation after they have succeeded in programming rudimentary approximations such as those presented here.

LUNAR CRATERS (DRAWINGS **97–99**)

Ever since Galileo first turned his crude telescope to look at the Moon, we have known that its surface is covered with circular depressions which we call *craters*, ranging in size from the vast lava-filled Maria Imbrium, which is a third as wide as the Moon itself, down to pits with a diameter of about 1 m. Nearly all of them are impact craters, rather than volcanic, and we must imagine a time when the Moon was under bombardment from chunks of matter of every conceivable size coming from outer space. Indeed, the entire Moon may be made up of matter which came together under gravitational attraction, so that what we are looking at on the Moon's surface are the scars of the last arrivals in this accretion process, more than 3000 million years ago.

Here on Earth the evidence of such early times has long since been lost in a surface which is continuously eroded by ice and water, and which has continued to churn with fresh mountain building and fresh ocean formation. But on Mars, Mercury and most of the known 33 moons which orbit the other planets that our space probes have visited we find a remarkably similar scene to that on our Moon: a surface more or less saturated with craters of all sizes, one on top of another in random fashion.

Although the locations of the craters are more or less uniformly random, the sizes of the craters appear to obey another rule. The best way to appreciate the nature of this rule is to consider the early *Ranger* probes to the Moon, which sent back pictures of the surface in quick succession while hurtling towards a crash landing. Each picture showed a magnified version of a small part of the preceding picture, as the probe got nearer to the Moon's surface. The pictures were remarkably similar in appearance, even though the first and last of the series showed widely different scales of view. The tiny craters were about as dense in their field as the large craters were in theirs. This is an example of *statistical self-similarity*.

To put this rather simply, there are a few large craters, more medium-sized craters and many more small craters. The approximate rule

97 Craters in the plane or spheres in space

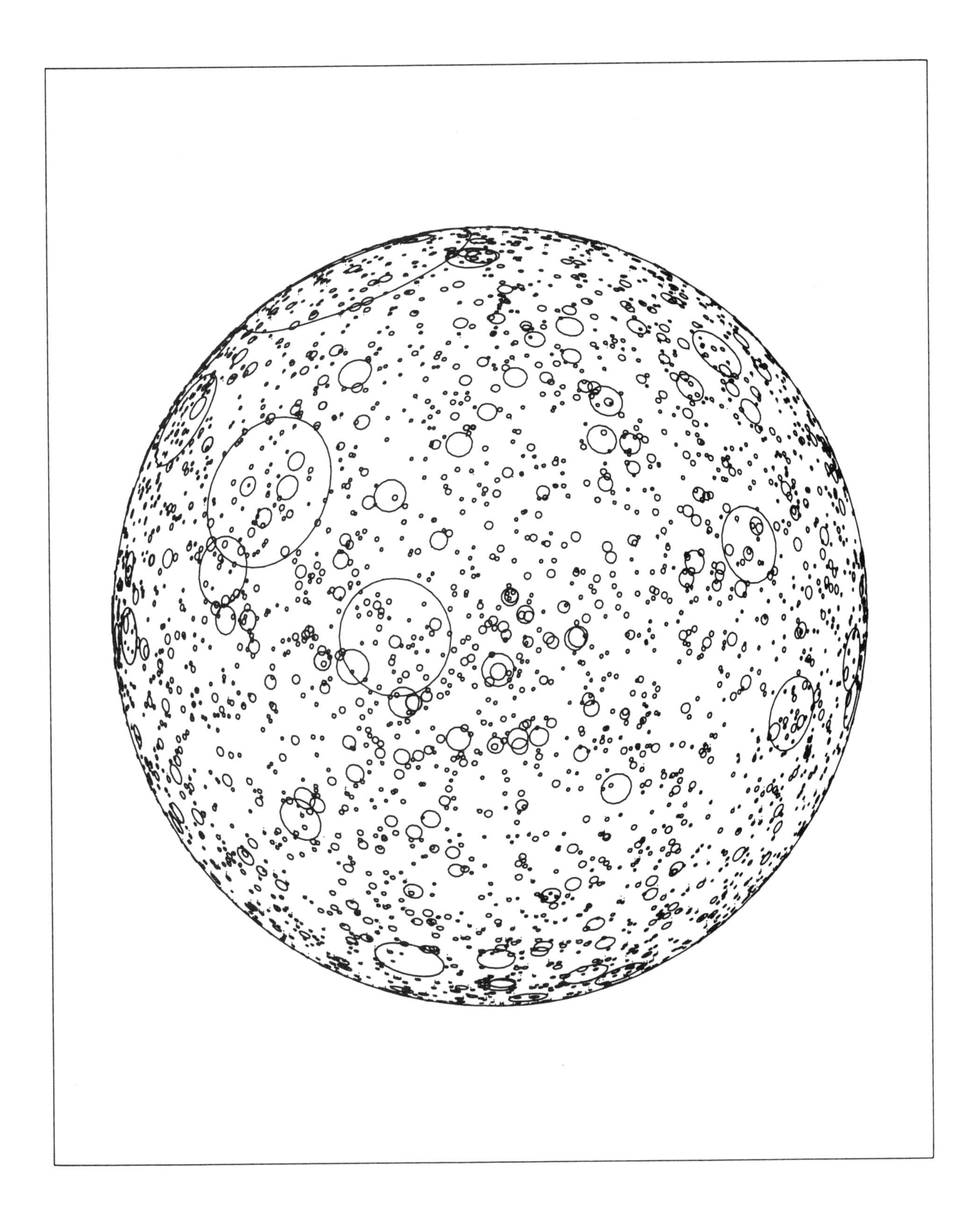

98 Moon

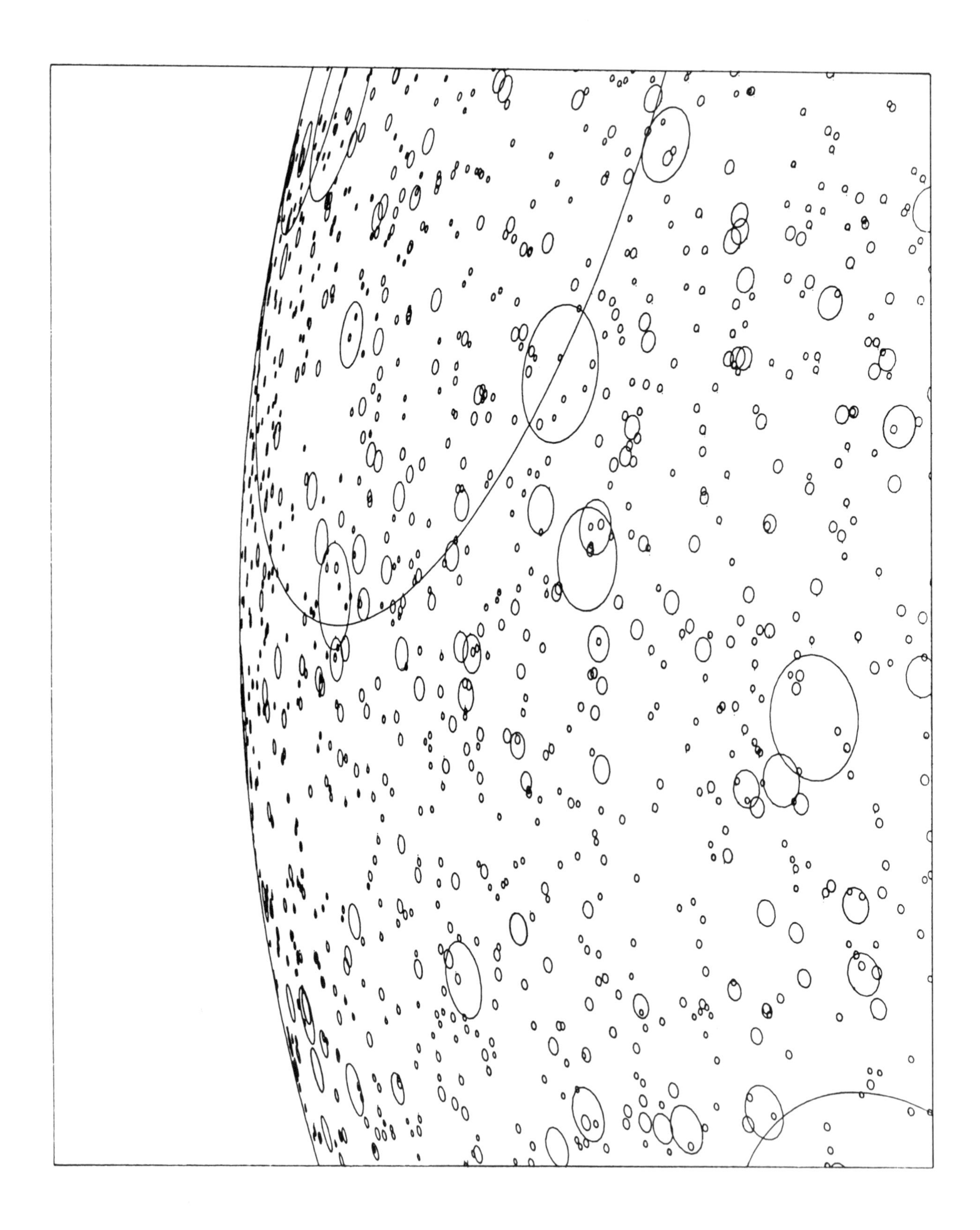

99 Moon view

for this may be stated thus: any given area of lunar surface is equally covered by all sizes of crater. It is necessary to be more precise than this by expressing the rule in probability terms:

$$\Pr(R \geqslant r) = r^{-2} \text{ for crater radii } 1 \leqslant r \leqslant \infty$$

In a computer program designed to simulate cratering, how shall we simulate this probability function using the uniform random function RND? Here is the chain of reasoning to solve this problem.

$$\Pr(\text{RND} \leqslant n) = n \text{ for } 0 \leqslant n \leqslant 1$$
$$\Pr(\text{RND}^{\frac{1}{2}} \leqslant n^{\frac{1}{2}}) = n$$
$$\Pr(\text{RND}^{-\frac{1}{2}} \geqslant n^{-\frac{1}{2}}) = n$$

By substituting $n = r^{-2}$, so that $n^{-\frac{1}{2}} = r$, we obtain

$$\Pr(\text{RND}^{-\frac{1}{2}} \geqslant r) = r^{-2} \text{ for } 1 \leqslant r \leqslant \infty$$

So the required function of RND to generate crater sizes is

RND ↑ (−1/2).

Having successfully programmed a simulation of craters randomly scattered in a flat rectangular area you should attempt the puzzle of programming the same pattern on the surface of the sphere. Imagine that you are looking at the pole of sphere aligned with the z axis pointing straight at you, with the x and y axes pointing in the usual directions. Having chosen a circle radius for the crater to be drawn, first imagine that the circle centre is placed at the pole facing you, and then that it is rotated twice: firstly about the x axis and secondly about the z axis. The rotation about the z axis must be a uniform random variable, but the rotation about the x axis must be non-uniform if the scattering on the spherical surface is to be uniform. What function of RND will work?

This problem is left for you to solve. Clue: what is the polar area enclosed by a circle of latitude $\theta°$ from the pole as a fraction of the area of a hemisphere?

When you have succeeded in answering this question, you will need to use a chain of probabilistic reasoning as above to obtain the appropriate function of RND for the x axis rotation.

One final remark should be made: if your circles representing craters are to lie flush with the surface of the sphere, you will need to centre them somewhat below the surface. The precise amount depends upon the size of the circle and presents a nice problem using Pythagoras' theorem.

WATER WAVES (DRAWINGS 100–106)

The model of water waves presented here must be regarded as a first approximation only. No attempt is made to represent the effects that a refraction by obstacles and shallows and by current flow and turbulence and that waves breaking have on wave propagation. For those patterns, you must go to the seaside and watch. Here we shall imagine that we are by a mill pond on a windless day which has been temporarily disturbed as, for example, by the impact of a pebble or by a pike snatching at its prey.

The disturbance to the horizontal water surface propagates as a sum of sine waves of various wavelengths and various amplitudes. They travel outwards from the source of disturbance in concentric circles. Each wavelength travels at its own speed; the longer the wavelength the faster it is. Speed is proportional to the square root of the wavelength. The

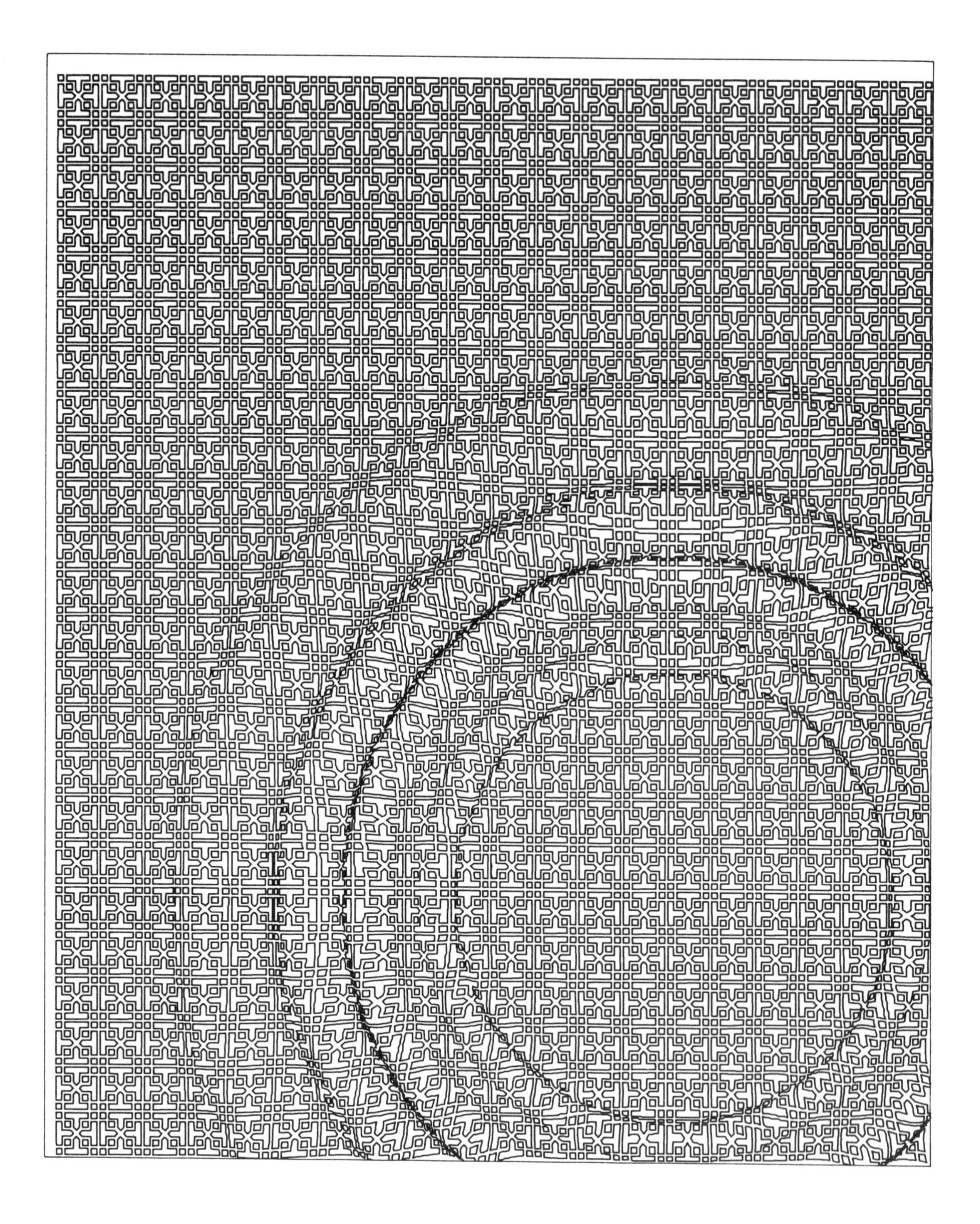

100 Ripple

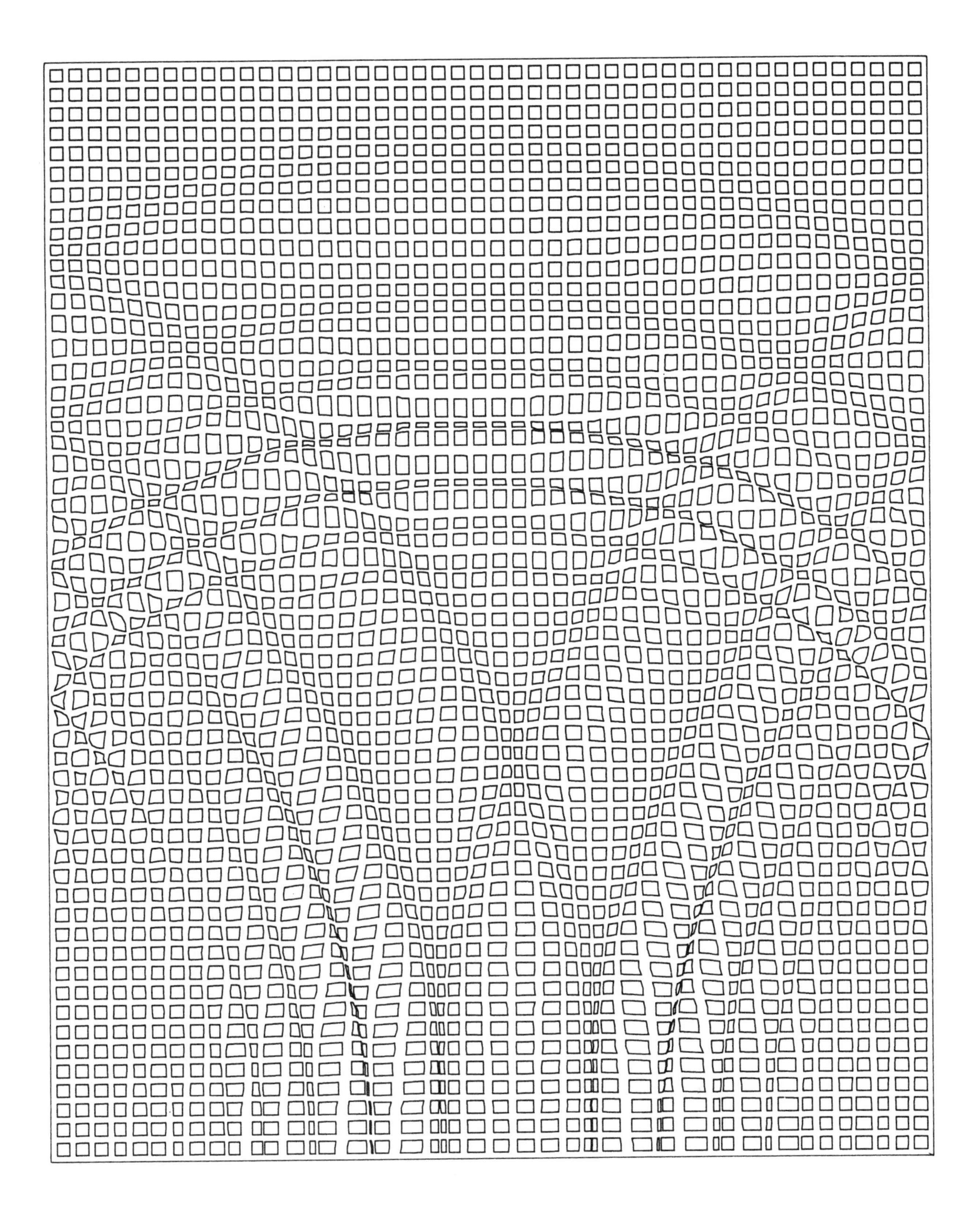

101 Ripple reflected

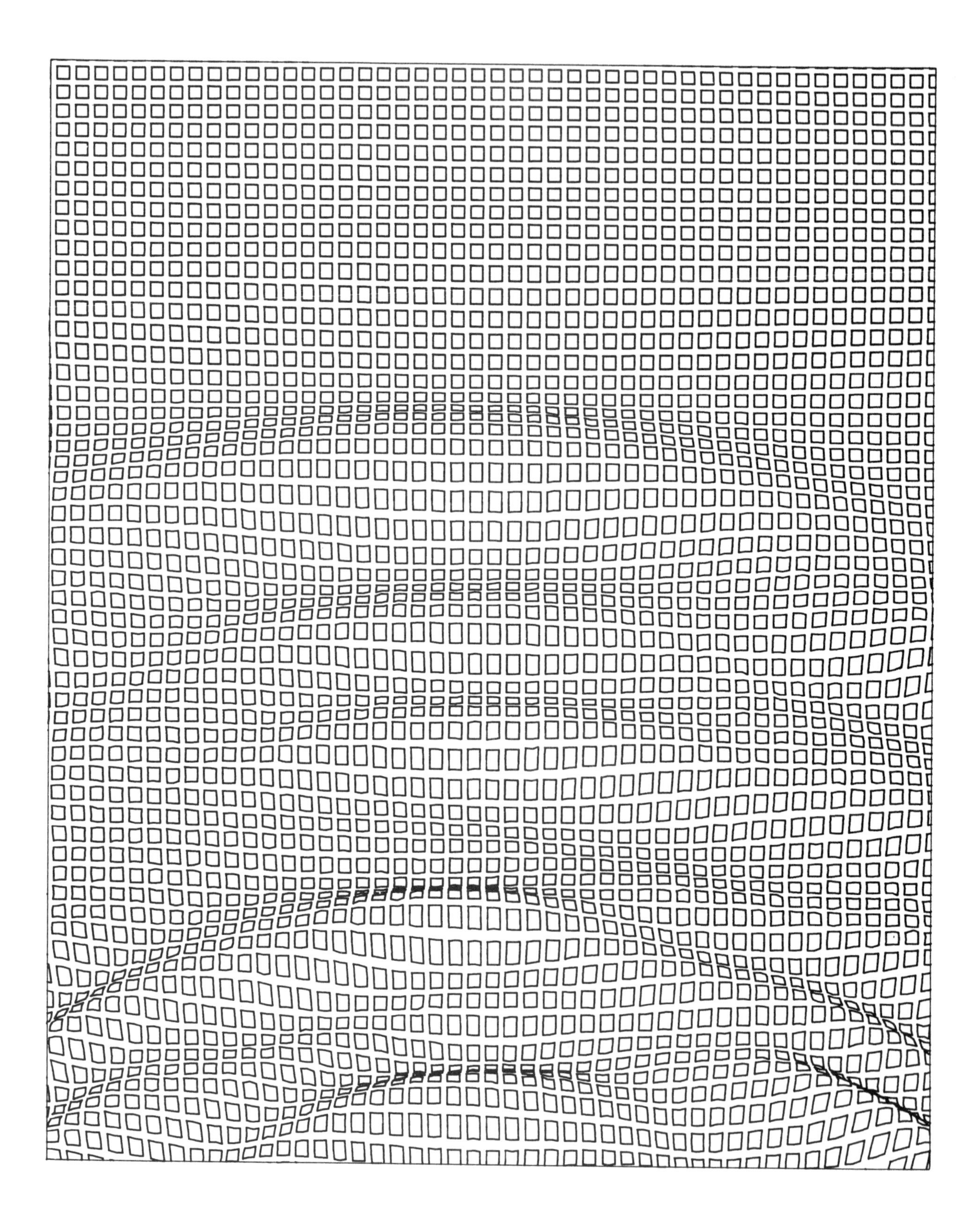

102 Ripples

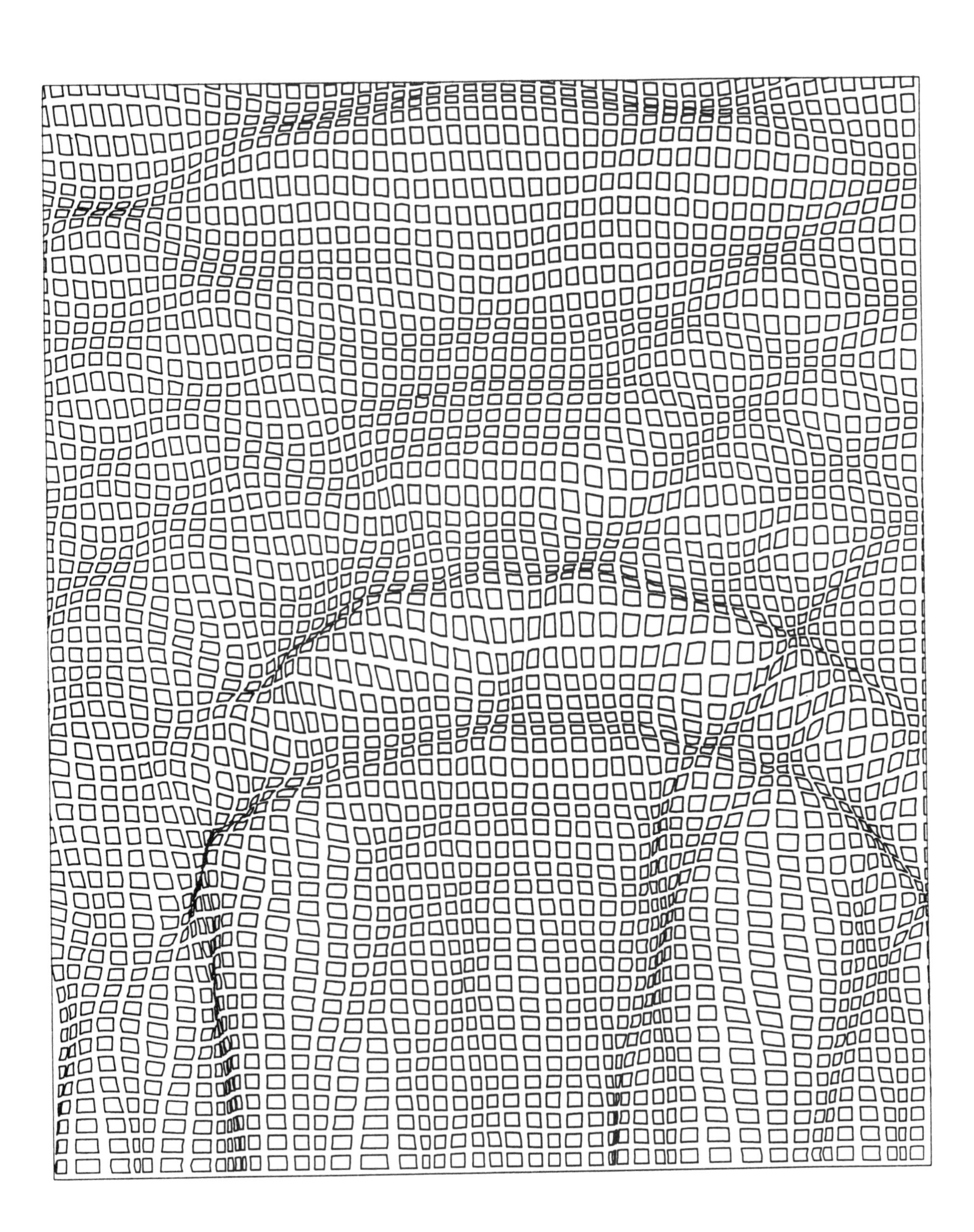

103 Ripples

individual component waves pass through each other, without effect on each other, except to form a sum displacement at any point:

$$y = a_1 \sin(b_1 x) + a_2 \sin(b_2 x) + a_3 \sin(b_3 x) + \ldots$$

a_n is the amplitude of the nth component, and $b_n = 2\pi/w_n$, where w_n is its wavelength. With so many combined choices to make for these program parameters, many different-looking wavy surfaces can be created.

It is important to realise that, although each wave component travels, the water itself does not travel, except to bob up and down on the spot, or to rotate in small circles or ellipses, by the amount of its wave amplitude. The energy of the original disturbance travels across the water acting as a medium. The disturbance, being temporary, provides a beginning and an end to the train of waves travelling outwards. Also, because different components travel at different speeds according to their wavelengths, *dispersion* takes place. That is why an initially abrupt disturbance eventually becomes spread out as a wave train.

The leading and trailing edges of a wave train of this kind are not abrupt. Both blend into the undisturbed water in a gradual manner. The wave train has an overall profile which is lens like in shape (Figure 5.4E). This effect is easily achieved by multiplying the sum of sines by the function for describing the lens profile.

There is one further wonder of our water worth pondering and worth watching for; the energy of this outwardly travelling wave packet travels at half the speed of the waves themselves! (It is rather like watching a tractor move at half the speed of its top tread.) You can see this quite clearly by selecting any wave crest in particular to watch as it starts up at the trailing edge, moves right through the lens profile until it reaches the leading edge of the wave train and disappears into the flatness ahead.

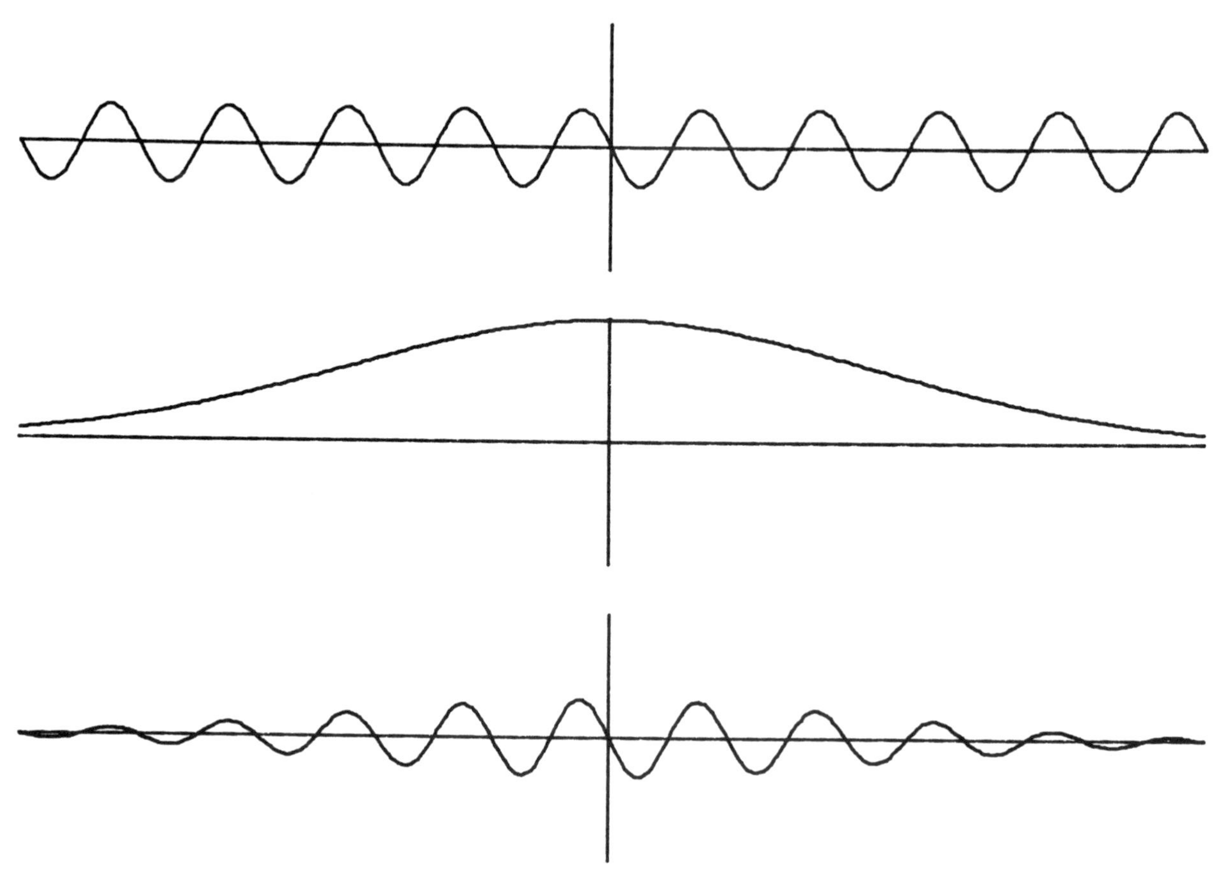

Figure 5.4E

Animators need to take account of this effect as well as that of dispersion, but it can be ignored in a still drawing.

It might be worth stressing how durable waves are. While the viscosity of water and air friction will eventually dissipate their energy so that the waves lose height as well as lengthen with age, the process is remarkably slow. Ocean waves would continue many more thousands of miles than they do if the land were not there to stop them. We can suppose them to be perpetual. We can also suppose that wave reflections, in a straight wall for example, are mathematically perfect.

Of course, as a concentric ripple expands from the centre of disturbance, its amplitude must decrease, not because of a loss of energy but because the same energy is shared out around a larger circle. Therefore the amplitude is inversely proportional to the square root of the ripple's radius. All these phenomena are shown in Drawing **101**.

Suppose that the originating disturbance is a long way off. Then, instead of using the model of concentric ripples which decay in amplitude with increasing radius, we model the waves as non-decaying and straight. We also dispense with the lens-like profile, assuming that dispersion has spread the wave train widely. The next interesting problem to consider is how to program wave trains travelling through each other from different directions. You may like to solve this and attempt to find gently irregular effects. The number of wave components and their range of wavelengths will be limited in practice by the computing time available to you, as sines are time-consuming calculations. The use of a *look-up array* becomes an essential time saver.

Reading

Beatty, O'Leary and Chaikin, 1981; Briggs and Taylor, 1982; Dixon, 1985b; Ferris, 1982; Kinsman, 1983; Mandelbrot, 1982; Smith, 1982; Stevens, 1977; Stewart, 1975; Tall, 1985.

104 Ripples

105 Waves

106 Turtle geometry

Bibliography

Aaboe, A., 1964, *Episodes from the Early History of Mathematics*, Random House, New York.
Abbott, P., 1940, *Trigonometry (Teach Yourself Series)*, English Universities Press, London.
Abbott, P., 1948, *Geometry (Teach Yourself Series)*, English Universities Press, London.
Ball, W.W.R., 1949, *Mathematical Recreations and Essays*, Macmillan, London.
Beatty, J.K., O'Leary, B., and Chaikin, A., 1981, *The New Solar System*, Cambridge University Press, Cambridge.
Boyer, C.B., 1968, *A History of Mathematics*, Wiley, New York.
Briggs, G.A., and Taylor, F.W., 1982, *The Cambridge Photographic Atlas of the Planets*, Cambridge University Press, Cambridge.
Cook, T., 1914, *The Curves of Life*, Dover Publications, New York.
Courant, R., 1934, *Differential and Integral Calculus*, Wiley, New York.
Courant, R., and Robbins, H., 1941, *What is Mathematics?*, Oxford University Press, Oxford.
Coxeter, H.S.M., 1961, *Introduction to Geometry*, Wiley, New York.
Critchlow, K., 1976, *Islamic Patterns*, Schoken Books, New York.
Critchlow, K., 1979, *Time Stands Still*, Gordon Fraser, London.
Cundy, H.M., and Rollett, A.R., 1961 (reprinted, Tarquin, 1981, Diss), *Mathematical Models*, Oxord University Press, Oxford.
Dixon, R., 1982, The drawing out of an egg, *New Scientist*, July 29.
Dixon, R., 1983a, Geometry comes up to date, *New Scientist*, May 5.
Dixon, R., 1983b, *The Mathematics and Computer Graphics of Spirals in Plants*, Leonardo, **XVI-2**.
Dixon, R., 1985a, The Fermat spiral, *MicroMath 1*(2).
Dixon, R., 1985b, Pentasnow, *Mathematics Teaching*, **110**.
Escher, M.C., 1972, *The Graphic Works of M.C. Escher*, Pan Books, London.
Ferris, T., 1982, *Galaxies*, Stewart, Tabori and Chang, New York.
Gombrich, E.H., 1964, *The Story of Art*, Phaidon, London.
Hardy, G.H., and Wright, E.M., 1938, *The Theory of Numbers*, Oxford University Press, Oxford.
Heath, T.L., 1956 (original, 1925); *Euclid — The Thirteen Books of The Elements*, Dover Publications, New York (3 vols).
Hilbert, D., and Cohn-Vossen, S., 1952, *Geometry and the Imagination*, Chelsea, New York.
Hogben, L., 1967, *Mathematics for the Million*, Pan Books, London.
Huntley, H.E., 1970, *The Divine Proportion*, Dover Publications, New York.
Kinsman, B., 1983 (original, 1965), *Wind Waves*, Dover Publications, New York.
Kline, M., 1972 (original, 1953), *Mathematics in Western Culture*, Pelican, London.
Kline, M., 1967, *Mathematics for the Non-Mathematician*, Dover Publications, New York.
Lamb, H., 1987, *Infinitesimal Calculus*, Cambridge University Press, Cambridge.
Lawlor, R., 1982, *Sacred Geometry*, Thames and Hudson, London.
Lawrence, J.D., 1972, *A Catalogue of Special Plane Curves*, Dover Publications, New York.
Lockwood, E.H., 1961, *A Book of Curves*, Cambridge University Press, Cambridge.
Mandelbrot, B.B., 1982, *The Fractal Geometry of Nature*, Freeman, San Francisco, California.
Markushevich, A.I., 1962, *Complex Numbers and Conformal Mappings*, Pergamon Press, Oxford.
Parramon, J.M., 1984, *Perspective — How to Draw*, Parramon Editions, Barcelona.
Pedoe, D., 1979, *Circles — A Mathematical View*, Dover Publications, New York.
Pedoe, D., 1976, *Geometry and the Liberal Arts*, Peregrine, London.
Resnikoff, H.L., and Wells, R.O., 1984, *Mathematics and Civilization*, Dover Publications, New York.
Roth, L., 1948, *Modern Elementary Geometry*, Nelson, London.
Smith, C.S., 1982, *Search for Structure*, Massachusetts Institute of Technology Press, Boston, Massachusetts.
Steinhaus, H., 1983 (original in Polish, 1938), *Mathematical Snapshots*, Oxford University Press, Oxford.
Stevens, P.S., 1977, *Patterns in Nature*, Peregrine, London.
Stevens, P.S., 1981, *A Handbook of Regular Patterns*, Massachusetts Institute of Technology Press, Boston, Massachusetts.
Stewart, I., 1975, *Concepts of Modern Mathematics*, Pelican, London.
Stewart, I., 1977, Gauss, *Scientific American*, July.
Tall, D., 1985, The gradient of a graph, *Mathematics Teaching*, **111**.
Thom, A., 1967, *Megalithic Sites in Britain*, Oxford University Press, Oxford.
Thompson, D'A.W., 1917, *On Growth and Form*, Cambridge University Press, Cambridge.
Vries, V. de, 1968, *Perspective*, Dover Publications, New York.
Weyl, H., 1982 (original 1952), *Symmetry*, Princeton, Boston, Massachusetts.

Answers

Exercise 4

7 **a** The line through their points of intersection.
b The perpendicular bisector of the line joining their centres.
c Another concentric circle of radius 4.

Exercise 5

3 **a** ∞. **b** Small. **c** No.

Exercise 7

1 **a** $\sqrt{26}$. **b** $2\sqrt{3}$. **c** $\sqrt{34}$.

2 **a** $\sqrt{5}$. **b** $\sqrt{3}$. **c** $\sqrt{2}$.

3 $\frac{1}{2}\sqrt{3}$.

4 **a** $2\sqrt{2}$. **b** $\frac{1}{2}\sqrt{15}$. **c** $\sqrt{2}$.

5 **a** $\sqrt{2}$. **b** $\sqrt{3}$.

6 $\sqrt{3}$.

7 $\frac{1}{2}\sqrt{2}$.

8 **a** $1:1:\sqrt{2}$. **b** 3:4:5. **c** $1:\sqrt{3}:2$.

Exercise 8

1 **a** Left-hand column: 11, 60, 61 and 13, 84, 85; given by the formula n, $\frac{1}{2}(n^2 - 1)$, $\frac{1}{2}(n^2 + 1)$, for all odd numbers $n > 1$.
b Right-hand column: 20, 99, 101 and 24, 143, 145; given by the formula $2n$, $n^2 - 1$, $n^2 + 1$, for all even numbers n.

Exercise 10

1 **a** 3.1410. **b** 3.1427.

Exercise 11

1 **a** $\left(\frac{14}{13} + \frac{14}{15}\right)$. **b** $+\frac{1}{13}$.

Exercise 12

1 **a** -0.174. **b** 0.479. **c** 0.707.

5 **a** PM. **b** OM. **c** ST. **d** OV. **e** OS. **f** UV.

6 **a** $\sqrt{(1 - x^2)}$. **b** $x/\sqrt{(1 - x^2)}$. **c** $1/x$.
d $1/\sqrt{(1 - x^2)}$. **e** $\{\sqrt{(1 - x^2)}\}/x$.

7 0.5 and 0.866.

8 0.866.

9 0.776 and 2.898.

10 221.7 m.

11 **a** 1.176. **b** 0.868. **c** 0.765. **d** 0.618. **e** 1.414. **f** 1. **g** 1.732.

12 **a** 41.3°. **b** 48.7°. **c** −20.49°. **d** Trick question, no answer.

13 **a** 22.62° and 67.38°. **b** 30° and 60°. **c** 26.57° and 63.43°.

14 54.74°.

15 **a** 45°. **b** 45°. **c** 54.74°. **d** 35.26°.

Exercise 13

2 $A^{10}/10!$ and $A^{11}/11!$

Exercise 14

1 **a** 57.296. **b** 180. **c** 343.775. **d** 90. **e** 45. **f** 120. **g** 72. **h** 36.

2 **a** 6.283. **b** 0.017. **c** 0.175. **d** 1.571. **e** 3.142. **f** 1.047. **g** 0.785. **h** 0.524.

3 R=SQR(X↑2+Y↑2) : A = ATN(Y/X) : IF X < 0 THEN A = A + 180 or A = ACS(X/R) : IF Y < 0 THEN A = 360 − A.

4 **a** A square. **b** A regular pentagon.

5 The angle whose cosine is.

Exercise 17

1 **a** The same mirror.
b Parallel mirrors.
c Intersecting mirrors.
d Never.
e Perpendicular mirrors.

2 **a** Same centre with equal but opposite angles of rotation.
b Equal but opposite angles of rotation about distinct centres.
c Otherwise.
d Never.

3 A scaling (dilatation) of factor $(R_2/R_1)^2$ about their common centre.

4 The circle $R = 0$.

5 **a** Yes.
b No, but more and more nearly so for smaller and smaller circles.

6 **a** A logarithmic spiral.
b A helix.

7 **a** X1 = Y
Y1 = X
b X1 = X * COS(2 * A) + Y * SIN(2 * A)
Y1 = X * SIN(2 * A) − Y * COS(2 * A).

8 $\sqrt{(A^2 + B^2 + C^2)}$.

Exercise 18

1
```
REM...BLANCMANGE CURVE
REM...BASE-WIDTH OF L AND LIFT-FACTOR K
INPUT L, K
DIM Y(1024)
J=1024
FOR N=1 TO 10
H=J*K
FOR M=0 TO 1023 STEP J
Y(M+J/2)=H+(Y(M)+Y(M+J))/2
NEXT M
J=J/2
NEXT N
REM...DRAWING COMMENCES
MOVE 0,0
FOR S=1 TO 1024
X=L*S/1024 : Y=Y(S)
DRAW X,Y
NEXT S
```

2 The Koch curve is programmed very like the blancmange curve, by computing an array of coordinates, but this time you need to compute *x* coordinates together with *y* coordinates. The problem on p. 176 is solved thus

```
X3=(X1+X2)/2 : Y3=(Y1+Y2)/2
X3=X3+K*(Y2-Y1) : Y3=Y3-K*(X2-X1)
```

for clockwise displacement, and

```
X3=X3-K*(Y2-Y1) : Y3=Y3+K*(X2-X1)
```

for anticlockwise displacement.

3 Dirichlet's function is given by

```
FOR N=2 TO 100
Y=H/N
FOR M=1 TO N-1
X=H*M/N
MOVE X,0 : DRAW X,Y
NEXT M
NEXT N
```

The subroutine which can be inserted into the above program to avoid repeated calls of the same fraction is as follows:

```
K=0
FOR S=2 TO M/2
IF N/S=INT(N/S) AND M/S=INT(M/S) THEN K=1
NEXT S
```

For flag $K=0$ draw the fraction M/N; for $K=1$, do not draw.

Index

Aaboe, A., 30, 32, 89
Abbott, P., 32, 61, 89
ABS, 109
accidental, 34
accretion, 93
affinity, 152
amplitude, 202
angle,
  constructable, 14–15
  in a semicircle, 29
  of view, 83, 144
angles, sum of in a triangle, 27
approximation, 34–51, 96
arbitrary, 1
arc, 2, 100
arccos, 103, 109, 110
Archimedes, 50, 89, 98, 100
Archimedes' spiral, 126, 138
architecture, 79, 170
arcsin, 103, 109
arctan, 109, 123
array, 175
  look-up, 203
art, 1, 170
auxiliary triangle, 56
axioms, Euclid's, 26–27

Babylonians, 14, 94, 96, 98, 99
balance, 4
Ball, W.W.R., 53
balloon, 4
BASIC, 107–109, 147
  mathematical functions, 109
Beatty, J.K., 203
bisection, 175
  of angle, 12, 99
  of line, 12
bishop's crook, 138
blancmange curve, 175
blotching, 1
bombardment, 193
Boyer, C.B., 89
Briggs, G.A., 203
brittle, 4
Brouncker, William, 101
Brunelleschi, Filippo, 79

calculator, pocket, 1, 34, 104
capillary action, 1
cardioid, 138
Cartesian coordinates, 123
  curves with, 122–123
casting, electrochemical, 4
Cayley's sextic, 138
centimetre, 2
centroid, 55
Chaikin, A., 203
chamfering, 1, 77
chord of a circle, 28, 98, 102
circle, 1, 12, 138
  area of, 99
  coaxial, 68–72
  compass-drawn, 1
  computer-drawn, 126
  mid-, 66–68
circumcentre, 55, 57
circumcircle, 55
circumference, 98
cissoid of Diocles, 138
clouds, 169
coastline, 169
cochleoid, 138
commensurable, 185
compass drawing, 1
compass, constructions using, 12
composition, 169
concentric circles, 197
conchoid of Nicomedes, 138
condensation, 186
conformal mapping, 160
congruent transformation, 147–152
conics, 139
construction, Euclidean, 12–32
continued fraction, 101
continuity, 186
coordinates,
  Cartesian, 107–108
  polar, 107–108, 126–138
cord, stretched, 1
corners, 175
cosec, 104
cosine, 102–106, 109
  computation of, 105–106
cosine rule, 105
cosine table, 102
Courant, R., 78, 96
Coxeter, H.S.M., 54, 61, 73, 168
crash, program, 123
craters, lunar, 169, 193—197
Critchlow, K., 11, 50
Cundy, H.M., 11, 78, 93, 139
Cundy and Rollett's egg, 11, 93
curvature,
  radius of, 4, 78
curve,
  as sum of circular arcs, 3–11
  blancmange, 175
  Cartesian, 122
  exponential, 123
  fractal, 170
  Koch, 175
  logistic, 123
  in perspective, 86
  polar, 138
  rose, 138
  smooth, 3
cylinderwrap, 167

daisy, 122, 131
  false, 138
  true, 138
decagon, regular, 31, 104
degrees, 108
derivative, 175
diagonal, 93
diameter, 99
DIM, 107
Dirichlet's function, 177
discontinuity, 138, 185
dispersion, 202

dissipate, 203
distance, 1
disturbance, 202
divergence, 131
division,
 arithmetic, 14
 geometric, 12
Dixon, R., 11, 73, 143, 203
DRAW, 108
DRAWP, 108, 126
droplet, 4, 186
Dürer, Albrecht, 3, 87, 186
dust, fractal, 170

e = 2.718..., 30
earth, 193
 crust of, 169
Easter, 3
eccentricity, 139
egg,
 bird's, 3–4
 Cundy and Rollett's, 11, 93
 Euclidean, 4, 11
 Moss', 93
 string-drawn, 75, 77
 Thom's, 93
Egyptians, 98
ellipse, 4, 77, 139
ellipsoid, 4
elongation, 4
energy, 203
enneagon, regular, 35, 40
envelope, 77
Eostre, 3
errors, 2
Escher, M.C., 88
Euclid, 3, 12–132, 89
Euler line, 51
evolute, 75–77
excircles, 58
EXP, 109
explanations, 26
exponential curve, 123

family of curves, 122
feedback, 107
Fermat primes, 53
Fermat's spiral, 126
Ferris, T., 203
Feuerbach's theorem, 59
Fibonacci numbers, 131
fingers, 1
fleas, 169
flexible, 4
fluke, 34
flush, 197
FOR...NEXT, 107
formula,
 angle-sum, 105–106
 perspective, 144
 polar-to-Cartesian conversion, 108
 radian-to-degree conversion, 123
fractal, 170
 artistic, 170
 curve, 170
 dust, 170
 geometric, 170–203
 surface, 170
fraction,
 continued, 101
 whole number, 14, 96, 177
Francesca, Piero della, 87
Freeth's nephroid, 138

galaxies, 169
Galileo, 193
Gauss, K.F., 52–53
glass blowing, 4
golden ratio, 12, 30, 50, 131
golden rectangle, 30
golden triangle, 30, 186
Gombrich, E.H., 88
GOSUB...RETURN, 107
gravity, 4, 193
Greeks, 34
Greenland, 167
gridding up, 86
growth, curve of, 123

Hardy, G.H., 53
Heath, T.L., 32
helical staircase, 80
heptagon, regular, 35–40, 104
hexagon, regular, 28, 93, 98
Hipparchus, 89, 105
Hogben, L., 89
horizon, 80
Huntley, H.E., 32
hyperbola, 139
hypotenuse, 92, 103–104

IF...THEN..., 107
Imbrium, Maria, 193
impossible constructions, 34–51
incentre, 58
inch, 2
incircle, 58
incommensurate, 97
inlets, 169
INT, 109
interactive, 107
intersection, 2
inverse points, 62
inversion, 62–65, 152
involute, 75–77
irrational, 97, 100, 177
islands, 169
isometry, 147–152

Kinsman, B., 203
Kline, M., 88
Koch curve, 175

lakes, 169
Lamb, H., 78
lamplight, 86
Lawlor, R., 50
Lawrence, J.D., 139
length, 2, 12
Leibnitz, Gottfried Wilhelm von, 101
Leonardo da Vinci, 88, 87
Leonardo's paradox, 83
limiting points, 72
line,
 straight, 1–2, 12, 79
 of sight, 1, 79
lituus, 138
Lockwood, E.H., 78, 139, 168
logistic curve, 123
look-up array, 203
loops, program, 122
lunar craters, 169, 193–197

Mandelbrot, Benoit B., 170, 175, 203
Mantegna, Andrea, 87
map, 167
Maria Imbrium, 193
mars, 193
Masaccio, 87
median, 55
Mercator, 160, 167
mercury, 193
mid-circles, 66–68
minimal surface, 4
mirror, 86, 150
mist, 169
model, 197
monotonic function, 126

moons, 169, 193
Moss' egg, 93
mountains, 169
MOVE, 108
MOVEP, 108

Nephroid, Freeth's, 138
Nicomedes, conchoid of, 138
nine, 15, 34
nine-point centre, 57
nine-point circle, 57
numbers,
  Fibonacci, 131
  whole, 14, 98

octagon, regular, 104
O'Leary, B., 203
orthocentre, 57
orthogonal circles, 65
orthogonal projection, 144

parabola, 139
parallel lines, 27, 80, 144
Parramon, J., 88
patterns, 169
pebbles, 169
Pedoe, D., 11, 61, 73, 88
pentagon, regular, 31, 104, 186
pentagram, 31
pentasnow, 186
perspective, 79–88, 144
pi (π) = 3.141 59..., 14, 30, 34, 44, 89, 98–101, 105
picture, 169
picture frame, 82
picture plane, 79
pins, 75
plane, picture, 79
plants, 131
plotter, 107
point, 2
  compass, 2
  vanishing, 80
polar coordinates, 108
pole of spiral, 126
polygon,
  regular, 98, 126
  star, 126
potter's wheel, 4
probability, 186–197
program language, 107
projection,
  parallel, 144
  point, 144
  stereographic, 167
promontories, 169
Ptolemy, 89, 98, 105
pyramid, 93, 105
Pythagoras, 94, 177
  theorem of: see theorem, of Pythagoras

radian, 99–100, 108
  conversion of to degrees, 108, 123
radical axis, 68
radical centre, 72
radius, 2, 99–100
  of curvature, 77
random, 109, 186
ratio, 12
  golden, 12, 30, 50, 131
rational, 185
ray, 79
realism, 193
rectifying the circle, 45
reflection, 86, 150
reiteration, 97
Resnikoff, H.L., 89, 168
revolution,
  surface of, 3
  solid of, 4
Richardson, L.F., 169
right angle, 28
rigid, 1
ripple, 169, 203
river, 169
RND, 109, 186, 197
Robbins, H., 89
rock, 169
Rollett, A.R., 11, 78, 93, 139
root,
  cube, 14
  square, 12, 14, 30, 96–97, 109
rotation, 147–150
Roth, L., 32, 61
rough, 176
rule, plastic, wooden or metal, 1

satellite, 169
scaling, 122, 152
screen dump, 107
screw thread, 1
sec, 104
self-projectivity, 186
self-similarity, 175
  statistical, 193
sequence, 97
series, 100–101, 106
seven, 15, 34
17-gon, 52–53
shadows, 86
shirt button, 77
short-cuts, construction, 13
sigmoidal, 123
sine, 102–106, 109, 123
  computation of, 105
sine rule, 105
sine table, 102
sine wave, 103
six-bar gate story, 83
sky, 169
slope, 169
Smith, C.S., 203
solid of revolution, 4
solidity, 4
space probe, 193
spin, 4
spiral, 126
  Archimedes', 126, 138
  equiangular, 126, 138
  Fermat's, 126, 138
  hyperbola, 138
  logarithmic (see spiral, equiangular)
  secondary, 131
springtime, 3
SQR, 109
square, 104
  perfect, 96
  in perspective, 85
square root, 12, 30, 96–97, 109
squaring the circle, 35, 44–49
star polygon, 126
stars, 169
Steiner, Jacob, 62
stereographic projection, 167
Stevens, P.S., 139, 168, 203
Stewart, I., 54, 186, 203
straight line, 1
stretched cord, 1
string drawing, 75–78
sunlight, 86
surface,
  fractal, 170
  minimal, 4
  of revolution, 3
Swift, Jonathan, 169
symbols, BASIC, 109
symmetry, 150

table,
cosine, 102
look-up, 203
sine, 102
Tall, D., 175, 203
tan, 103
tangency, point of, 13
tangent, 13, 28
tau (τ) = 1.618 034 ..., 30, 131
template, 77
texture, 169
theorem, 26
of Pythagoras, 31, 35, 92–95, 103, 197
proofs of, 95
Thom, A., 11, 93
Thom's egg, 93
Thompson, D'A.W., 78
three dimensions, 79
transcendental, 100
transformation, geometric, 147–168
translation, 122, 147–150
transversal, 27
trees, 169
triangle,
equilateral, 28, 93, 104
golden, 186
isosceles, 28, 93
right-angled, 28, 92–95, 103
triple coordinates, 144
triplets, Pythagorean, 94
trisection of an angle, 15, 34, 50
turbulence, 169

Uccello, Paolo, 87
uniform random variable, 186

Van Eyck, Jan, 87
variable, random, 109
vector graphics, 107
viscosity, 169
Viete, 100
view, angle of, 83, 144
viewpoint, 144
Vries, V. de, 88

Wallis, John, 101
warp, 1
water, 79, 197
wave, 123, 197
ocean, 203
water, 197–203
wave component, 202
wave profile, 202
wave train, 202
wavelength, 202
wear, 1
Wells, R.O., 89, 168
Weyl, H., 168
whole numbers, 98
whorls, 169
wiggliness, 176
Wright, E.M., 53

yin yang symbol, 11

Zeeman's paradox, 82